About Demos

Who we are

Demos is the think tank for everyday democracy. We believe everyone should be able to make personal choices in their daily lives that contribute to the common good. Our aim is to put this democratic idea into practice by working with organisations in ways that make them more effective and legitimate.

What we work on

We focus on six areas: public services; science and technology; cities and public space; people and communities; arts and culture; and global security.

Who we work with

Our partners include policy-makers, companies, public service providers and social entrepreneurs. Demos is not linked to any party but we work with politicians across political divides. Our international network – which extends across Eastern Europe, Scandinavia, Australia, Brazil, India and China – provides a global perspective and enables us to work across borders.

How we work

Demos knows the importance of learning from experience. We test and improve our ideas in practice by working with people who can make change happen. Our collaborative approach means that our partners share in the creation and ownership of new ideas.

What we offer

We analyse social and political change, which we connect to innovation and learning in organisations. We help our partners show thought leadership and respond to emerging policy challenges.

How we communicate

As an independent voice, we can create debates that lead to real change. We use the media, public events, workshops and publications to communicate our ideas. All our books can be downloaded free from the Demos website.

www.demos.co.uk

First published in 2006
© Demos
Some rights reserved – see copyright licence for details

ISBN 1 84180 156 9
Copy edited by Julie Pickard
Typeset by utimestwo, Collingtree, Northants
Printed by Calverts, London

For further information and
subscription details please contact:

Demos
Magdalen House
136 Tooley Street
London SE1 2TU

telephone: 0845 458 5949
email: hello@demos.co.uk
web: www.demos.co.uk

Governing at the Nanoscale

People, policies and emerging technologies

Matthew Kearnes
Phil Macnaghten
James Wilsdon

DEM⊘S

Contents

Acknowledgements

The research that led to this pamphlet was generously supported by the UK's Economic and Social Research Council (ESRC), through an award from its Sustainable Technologies Programme. We are grateful to the ESRC, and particularly to Fred Steward, Adrian Monaghan and Franz Berkhout for their leadership of that programme.

The project, which began in January 2004, has been a genuine team effort, and we have relied at every stage on the invaluable contributions of our Lancaster University and Demos colleagues: Robin Grove-White, Brian Wynne, Paul Miller and Jack Stilgoe.

Special thanks must also go to Rob Doubleday and Mark Welland at the Cambridge Nanoscience Centre for their support, and to George Smith, for allowing us to spend time in his labs in Oxford.

Over the course of the project, we have benefited from discussions with the growing community of thinkers and policy-makers working in this field. In particular, we would like to acknowledge the insights of Brian Bloomfield, Adrian Butt, Sheila Jasanoff, Richard Jones, Gary Kass, Les Levidow, Alfred Nordmann, Arie Rip, Dan Sarewitz, Andy Stirling, Bron Szerszynski, Jim Thomas, John Urry, Claire Waterton and Richard Wilson. Initial meetings of the International Nanotechnology and Society Network (www.nanoandsociety.com) have also provided thoughtful opportunities for exchange.

Thanks also to Johannes Vogel, Natalie Bell and Sarah Hoyle at the Natural History Museum for hosting our final workshop. Our ability to communicate the results of the project has been greatly enhanced by the work of Hugh Hartford, who has produced a film based on that workshop, and the artist Tim Caswell, who captured the tone of the discussion on a giant mural.

Finally, we are grateful to everyone who took part in the research: the scientists and experts we interviewed, and the members of the public who participated in the focus groups.

Matthew Kearnes
Phil Macnaghten
James Wilsdon
April 2006

About the authors

Matthew Kearnes is a research fellow in the Sociology Department at Lancaster University, where his research explores the social and cultural dynamics of science. He was the lead researcher on the ESRC-funded project 'Nanotechnology, risk and sustainability: moving public engagement upstream' and is also involved in the 'NanoDialogues' project that is being funded by the UK government's Sciencewise programme. Previously, he was a research fellow in the Geography Department at the Open University, where he looked at urban green spaces and conservation practices. He is a founding member of the International Nanotechnology and Society Network and a member of UK Nanotechnologies Standardization Committee.

Phil Macnaghten is a senior lecturer in the Sociology Department at Lancaster University, and from June 2006 will be Professor of Geography and Director of the Institute for Hazard and Risk Research at Durham University. His research centres on the cultural dimensions of environmental and innovation policy and their intersection with everyday practice. He has published widely on the social science of technology and the environment and is author, with John Urry, of *Contested Natures* (Sage, 1998). He advises the UK Economic and Social Research Board on nanotechnology and society, is a founding member of the International Nanotechnology and Society Network, and a member of the UK Nanotechnologies Standardization Committee.

James Wilsdon is head of science and innovation at Demos and a senior research fellow in the Institute for Advanced Studies at Lancaster University. His work focuses on the relationship between science, technology, democracy and sustainability, and his recent publications include *Better Humans? The politics of human enhancement and life extension* (edited with Paul Miller, Demos, 2006); *The Public Value of Science* (with Brian Wynne and Jack Stilgoe, Demos, 2005); *See-through Science: Why public engagement needs to move upstream* (with Rebecca Willis, Demos, 2004); and *The Adaptive State* (edited with Tom Bentley, Demos, 2003). He writes regularly for the *Financial Times*, and advises government and a wide range of organisations on science policy.

1. Behind the scenes at the museum

It was a typical Saturday morning at the Natural History Museum. Hordes of kids swarmed through the museum's doors, yelling and pointing as they caught their first glimpse of the diplodocus, whose 26-metre skeleton takes pride of place in the cavernous entrance hall.

Dotted among the crowd were a handful of people with something far smaller on their minds. Picking their way past the diplodocus, up the stairs, down a corridor, and through an unmarked door, these invited guests assembled in the museum's board room. Over coffee and biscuits, they started talking. Half of the group were scientists: six PhD students, three junior researchers, two lecturers and one professor. The other half were members of the general public: parents, teachers, an osteopath, someone in IT. Some had travelled down from Manchester, others from Enfield in north London. But despite their varied backgrounds, they had come together to discuss one thing: the social and ethical implications of nanotechnologies.

The meeting at the museum, on 5 November 2005, was the culmination of a two-year project carried out by Lancaster University and Demos. Through a focus on nanotechnologies, the project sought to develop opportunities for 'upstream' dialogue between scientists and citizens, which would enable public voices to be heard at a stage when they could still influence research priorities.

The politics of small things

The starting point for the project was a recognition that policy and media debate about nanotechnologies is gradually intensifying. A lot of the discussion so far has been characterised by competing visions of promise and threat. For its cheerleaders, nanotechnology is seen to be ushering in a 'new industrial revolution' that will include break-throughs in computer efficiency, pharmaceuticals, nerve and tissue repair, catalysts, sensors, telecommunications and pollution control.[1] Research funding for nanotechnologies has increased exponentially, with Europe, the US and Japan each spending more than US$1.2 billion in 2005, up from around $100 million a decade ago.[2]

At the same time, social and environmental concerns, which originated with dystopian predictions of 'grey goo', have taken on a sharper focus around the potential toxicity of nanoparticles and the need for tighter regulation.[3] An influential report by the Royal Society and Royal Academy of Engineering (RAE), published in July 2004, provided a more nuanced account of benefits, risks and uncertainties. And in an attempt to learn from earlier controversies, the report called for 'a constructive and proactive debate about the future of nanotechnologies [to] be undertaken now – at a stage when it can inform key decisions about their development and before deeply entrenched or polarised positions appear'.[4]

This commitment to 'upstream' public engagement raises many unresolved questions. At what stages in scientific research is it realistic to raise issues of public accountability and social concern? How and on whose terms should such issues be debated? Are dominant frameworks of risk, ethics and regulation adequate? Can citizens exercise any meaningful influence over the pace, direction and interactions between technological and social change? How can engagement be reconciled with the need to maintain the inde-pendence of science, and the economic dynamism of its applications?

The Lancaster–Demos project sought to answer these questions. This pamphlet summarises our findings, with each chapter corres-ponding to a different phase of our research.

In the UK, a standard reference point in discussions about nano-technologies, particularly in 2003 and 2004, was the controversy over genetically modified (GM) crops and foods, which was fresh in people's minds. Such a comparison cannot be applied in any straightforward way, but because of its prominence as an interpretive frame, we started our research by looking for lessons that could be drawn from early discussions around agricultural biotechnology (from the 1970s onwards) and applied to nanotechnologies today. We interviewed ten leading players in the GM debate and invited them to reflect back on initial trajectories of public, political and media concern.[5] We asked whether, with hindsight, they would have done anything differently, and what approaches they would recommend now in relation to nanotechnologies. Chapter 2 draws some common threads from these interviews, while also interrogating the bio–nano comparison.

From the outset, our research tried to embrace the complexities of nanotechnologies, and their emerging social dimensions. Part of the challenge was to understand the implicit assumptions, values and visions – what are sometimes termed the '*imaginaries*' – of key actors in the nanotechnology domain.[6] Imaginaries are projections of future imagined worlds embedded within the present, which frequently inform and shape new scientific fields.[7]

The second phase of our project sought to explore some of the imaginaries at work within nanotechnology through a further round of interviews with key opinion formers,[8] and a period of ethnography in nanoscience labs at Cambridge and Oxford Universities. Based on these encounters, chapter 3 identifies five 'programmatic imaginaries' that fuel expectations of the promise of nanotechnologies.

Next, we turned our attention to the views and concerns of the wider public. In London and Manchester, we facilitated a series of five focus groups, each of which met twice. The aim of this exercise was less to obtain a representative cross-section of opinion, and more about flushing out the kinds of issues (social, ethical, political, economic, scientific) that might spark controversy if debate about nanotech-nologies were to gather momentum in the UK. The focus groups gen-erated a rich collection of insights, which are summarised in chapter 4.

The final phase of the research brought us to the Natural History Museum, where our 12 nanoscientists (including some who had been interviewed as part of the lab-based ethnography) spent a day with our 12 members of the public (all of whom had participated in the focus groups). This was a surprisingly open and positive conversation, which generated a lot of common ground. Chapter 5 tells more of the story of that day.

The Lancaster–Demos project has been one – relatively modest – attempt to develop more meaningful forms of 'upstream' public engagement. A growing number of organisations and individual scientists are experimenting in similar ways. Most of these are embryonic processes, and their eventual results are still uncertain. But they are important sites of social learning, and in the final chapter we also draw out some implications of our research for future processes of public engagement. This is followed by an afterword by Brian Wynne, who offers some insights from the project for social science.

Behind the scenes at the museum, we wanted to lay a few dinosaurs to rest. Contrary to those who insist that ordinary people are incapable of having a productive dialogue with scientists about complex technological questions,[9] we hoped to demonstrate that a productive exchange is possible on 'upstream' questions and uncertainties. In this respect, the meeting was a success. Our lay participants left feeling empowered by their encounter with the scientists, and our scientists were grateful for the opportunity to reflect on the social dimensions of their research. As one nanoscientist remarked at the end of the day:

> I was interested to see the things that people kept coming back to . . . were the whole issue of responsibility and how we actually use technology. These are the same sort of issues we don't know anything about and have no control over. We possibly ought to. It's the same things I worry about when I'm not being a scientist. So in a way it's quite nice . . . I'm not as detached from the real world as I thought I was.

2. From bio to nano

Given the starkness of the 'GM controversy', particularly as it unfolded in Europe, it is not surprising that there has been speculation as to whether nanotechnologies might experience a similarly rough passage. Here is another potentially transformative technology, subject to similar levels of utopian promise, expectation and dystopian fear.[10] Crudely put, the GM experience represents a warning, a cautionary tale of how not to allay public concern. Avoiding nanotechnology becoming 'the next GM' is seen as critical to the public acceptability of applications in the field.[11]

Under scrutiny, the GM–nano analogy quickly breaks down. These are very different technical endeavours, emanating from different disciplines. One is a particular type of application, the other a catch-all for a multitude of products and processes. So a direct comparison between them is of limited value. We agree with the authors of one recent paper that the analogy 'is not as strong or as helpful as its ubiquity would suggest . . . [and] therefore needs to be employed advisedly'.[12]

But as we have suggested elsewhere, the GM case can still be useful to illustrate how policy-makers struggle to handle emerging technologies in the early stage of their development.[13] There are also various ways in which the GM experience has shaped, and will continue to shape, political and regulatory debates around nano-technologies. This chapter offers some critical reflection on a series of interviews with key individuals active in the pre-1999 debates over GM plants and crops in Europe.[14] Based on these interviews, we argue that there are useful lessons to draw from two sets of competing understandings in the GM controversy: *competing understandings of 'the science'* and *competing understandings of 'the public'.*

Competing understandings of 'the science'

In the 1970s many leading genetic scientists expressed effusive visions of the transformative societal futures that would result from advances in genetics and biology. One such figure, CH Waddington, described the arrival of genetics as presaging a 'second industrial revolution', which would overturn the destructive effects of the first revolution, which was based (in his view) on physics and chemistry.[15] Visions such as Waddington's were not simply *scientific* imaginaries. They were *social* too.

One of our interviewees, Professor Nigel Poole, articulated such an imaginary when he spoke with passion about the potential for genetic and plant science to transform the economy:

> *I remember so clearly getting a very passionate talk, a lecture, evangelical almost about the future of biotech. This must have been in the very early 1970s. And I was totally convinced – that in biotech we would start to see the end of the chemical industry or massive change in the chemical industry. And I think they even said that by the turn of the millennium the chemical industry would have been gone. . . . I don't really think then we were thinking about DNA, you know gene therapy and that stuff – that was a bit too early. But those were the dreams and that's still my belief. It's a belief that goes right back to 1972.*[16]

In the commercial sphere, Monsanto's initial R&D commitment to GM crops was justified in terms of equally positive visions for the future of global agriculture, beyond more technical visions of 'terminator technology' or proprietary brand herbicide resistance.[17] Although now often disparaged as having been focused exclusively on corporate profit and control, Monsanto's imaginaries in the 1980s and 1990s reflected a vision of a more environmentally benign system of food production. Equally striking, however, was the degree of naiveté within this vision about other actors' responses and expectations.

Societal and scientific imaginaries of this kind – projections of future imagined worlds – frequently inform and shape new scientific fields. The GM experience points to the fact that, despite their scientific significance and persuasive power for governments and investors, such imaginaries tend to be insulated from wider recognition, accountability and negotiation. They are shielded by myths about the purity of science and assumptions of a linear relationship between scientific research and the public domain.[18] According to this model, it is only when scientific knowledge is thought to have potential 'applications' that social and ethical dimensions enter in. This means that social issues are acknowledged to arise only in connection with possible impacts, not with the aims and purposes underlying scientific knowledge production.

However, in the last decade or more this model has come under increasingly intense pressure, partly due to the changing political economy of research where commercial exploitation and property rights have become central, and partly due to the emerging policy significance of 'public engagement' in the UK and EU. Under these conditions, the need has intensified for even 'basic' scientists to project images of how their research might benefit society in the future. As basic research comes to be called 'pre-market' research, an unavoidable implication is that 'basic' research practices are imagining possible market outcomes, in ways that may subtly but significantly shape those research agendas and cultures themselves.

The limits of risk assessment

The regulatory context for GM crops was framed by a particular conception of risk assessment – one which was methodologically quantitative and almost exclusively concerned with the 'direct' effects of individual crops. Wider questions arising from the overall social, ecological, medical and political implications of GM technology were marginal to official considerations. This limited framework of risk assessment, coupled with official assurances of safety, had the effect of making the official mechanism for risk assessment a de facto locus for the political contestation of GM releases.[19] It also played a role in the formation of public controversy in the late 1990s.

In the UK, the Environmental Protection Act 1990 established the Advisory Committee for Releases to the Environment (ACRE) as the formal body responsible for assessing the risks to human health and the environment from the release of GM organisms. ACRE's position was awkward from the outset. As the only established mechanism for the regulatory assessment of GM releases this *advisory* body became the de facto *political* authority on GM releases, backed by the government's commitment to 'sound science'.[20] However, ACRE was concerned solely with the risks of individual GM crops. In seeking to address specific risks on a case-by-case basis, this risk assessment template came to be structurally built on past knowledge, rather than taking account of the potential for new types of hazards that might arise in unknown forms.[21]

The ex-chair of ACRE confirmed in his interview with us the difficulties that this methodology created in relation to the wider cumulative implications of GM crops:

> *We recognised quite quickly in ACRE that it was really very easy to give approval, say, for GM maize as is being done at the moment. You could not see any human risks, you couldn't really see any serious environmental ones, and as was proven in the farm trials, it's actually slightly better than traditional herbicide treatment in terms of wildlife. But we asked the question, sure, we can do this for one crop, one manipulation. But when all crops are being manipulated, every effect becomes additive. So if you approve an insect-resistant oilseed rape, you can do an analysis and say, well, that particular variety is only likely to occupy such a percentage of the area of the UK. The impact on insect production is small, the impact on birds is therefore likely*

to be small, probably quite acceptable. . . . . However, if every farmer grew those crops at every farm, suddenly the impact is enormous. Where is the mechanism to put it all together?[22]

He expanded on this concern later in the interview:

The big issue in terms of commercialising is what happens if you then approve another variety with another gene and then another variety with another gene. You'd need to know something about the interrelationship of those genes if they come together. And I finished chairing the committee before it was properly decided. . . . First person's dead easy, second person has to take into consideration the first gene, the third has to take into consideration the first two, the fourth has then got three prior genes plus their own. So there were lots of arguments. I think it's still not remotely solved as to what happens when you've got lots of different genes out there.[23]

Though initially imagined in precautionary terms, ACRE's reductionist framing stunted the extent to which real-world contingencies could be thoroughly considered. This led to mounting problems for the authorities responsible for the regulation of biotechnologies. Importantly, the limited framework of risk assessment was also intimately linked to the marginalisation of wider social and ethical concerns about GM food.[24] Such concerns – including the perception that government decisions had already been taken, that GM foods would lead to an inevitable diminution in consumer choice, of GM as unnatural, and concerns about corporate control of food systems – were simply not captured by the language of risk and safety.[25]

The effect of this deletion was to make debates about the risk and safety of GM crops stand in for a host of other unacknowledged concerns.[26] Yet the poignancy of these wider social concerns was redoubled by the lack of any official recognition and official assurance of the adequacy of assessment mechanisms. And for these precise reasons, ACRE became the de facto locus for the political contestation of GM releases.

Other European governments, including those of Denmark, the Netherlands, Germany and Norway, responded to the concerns raised about GM with more innovative forms of social debate and dialogue. Building on these, two such initiatives were undertaken in the UK – a consensus conference, organised in 1995 by the Science Museum and what would become the Biotechnology and Biological Sciences Research Council (BBSRC), and a government-organised 'National Biotechnology Conference' held in early 1997.[27] Unfortunately, both of these initiatives were limited in their scope, public visibility and ability to shape the trajectory of GM regulation and development. Similarly, neither was framed to enable detailed examination of wider societal and ethical concerns.

Competing understandings of 'the public'

During the 1970s and 1980s, public attitudes to nuclear power were systematically characterised as subjective, emotional and false risk perceptions.[28] In the early 1990s, an equivalent dynamic emerged in the biotechnology field. With a few exceptions, it was assumed that public concerns about GM crops could be founded only on an incorrect understanding of the technology or a complete lack of knowledge altogether.

As the 1990s advanced, social science researchers became increasingly active observers of the state of public opinion in relation to GM plants and foods.[29] Much of this work focused on public attitudes rather than underlying sources of social tension, and how these reflected limitations in the risk–regulatory framework itself. Indeed, most built on the assumption that the discourse of atomised science-defined 'risks' offered an analytically sound basis for commentary on the state of public opinion. As such, even though survey data began to point to a steady decline of public confidence towards biotechnology throughout the 1990s, this provided little explanation or warning for why GM would become the focus of such controversy. The assumption was that the key issues of public concern were the risks as defined by risk assessment, and that any disinclination by the public to accept such risks was based on a (false) belief that the risks were too high.

Even following the official discrediting of this 'deficit model' (symbolically put to bed in the House of Lords *Science and Society* report in 2000[30]), this misconception continues to be resurrected, albeit in a succession of new versions. Such persistence reflects an institutional science and policy culture which continues to project problems of public conflict, mistrust and scepticism about prevailing science on to other supposedly blameworthy agents – often a sensationalist media, or mischievous non-governmental organisations (NGOs). Responsibility for such problems is continually externalised away from official institutions, such that governments' and scientists' own roles are rarely questioned.

Some of our interviewees reflected this view:

> *There was a clear view that there was an anti-science agenda that was coming through. . . . The biggest frustration was the dishonesty and the distortion [on the part of NGOs and the media], which it's very difficult to handle. It's extraordinarily difficult to handle.*[31]

> *Fear of the unknown . . . it's like MMR in many ways. You know, no real benefit – and fear of the consequences – and a confusion because they were being fed downright lies by people. There is no way of actually correcting the [NGO] lies.*[32]

The implication is that NGOs purposefully acted to manipulate policy and create controversy. Yet interviews with NGO actors involved with GM campaigns throughout the 1980s and 1990s suggest that such charges misrepresent the capacity of NGOs. National bodies like Greenpeace, Friends of the Earth, the Royal Society for the Protection of Birds (RSPB) and the Soil Association, each of which made a distinct contribution to the more visible stages of the GM controversy, tend to be preoccupied with multiple issues. In the UK, these NGOs were relatively slow and uneven in developing coherent campaigns on GM crops. Indeed, the overall response by these groups to GM lacked clarity and unanimity. Greenpeace, for example, following its initial direct action drawing attention to Monsanto's first shipment of GM soya in mid-1996 was uncertain what do next. There was protracted internal discussion within the UK office about whether there was any appropriate basis for further initiatives. Friends of the Earth took up the issue only in 1998, in parallel with the RSPB's shared concern over the specific issue of potential biodiversity impacts from commercial growing of GM crops. This led to the setting up of the government's farm-scale trials at the end of that year.

So, far from leading the mounting controversies about GM commercialisation up to this point, the NGOs found themselves in the position of responding to the intensity of wider public unease being expressed through the spontaneous emergence of new networks and initiatives.[33] Whatever the beliefs or inclinations of individual supporters or staff members, NGOs face constraints in their ability to influence or transmit the full range of concerns of the wider population in relation to new technologies. Much of the difficulty for Greenpeace, Friends of the Earth and others in campaigning coherently on GM-related issues arose from the fact that the dominant 'risk' discourse offered them minimal scope for interventions. For example, Greenpeace's stated approach to GM issues was articulated in the idioms of science alone:

The difficulty Greenpeace has, is that we are a global organisation and, if one is to take value-based stances on what is and is not natural and the value judgements and the sort of loadings that that comes with, how relevant is it to talk about it in those terms and try and explain one's concern in those terms in China, where the term for nature doesn't actually exist or certainly doesn't exist in any meaningful form that we would recognise in the West? . . . That is not our position. Our position is about scientific risks. Our kind of globally applicable standard is the science of environmental risk. You can say that's the basis of our campaign policy and that's where we're coming from.[34]

Condensation points

By the end of the 1990s, GM crops had become something of an iconic environmental and social issue in many countries. At the immediate level, concern crystallised around the potential for unforeseen ecological consequences and the implications of GM for agriculture and food production. But discussion of the technology also reflected a broader set of tensions: global drives towards new forms of proprietary knowledge; shifting patterns of ownership and control in the food chain; issues of corporate responsibility and corporate proximity to governments; intensifying relationships of science to the worlds of power and commerce; unease about hubristic approaches to limits in human understanding; and conflicting interpretations of what might be meant by sustainable development. These and numerous other 'non-scientific' issues condensed on to GM crops because of a particular range of institutional and cultural contingencies shaping the technology and its development.[35]

This was hardly without precedent. In the very different circumstances of the 1970s, disputes about civil nuclear power had played something of an analogous role. Here too was an apparently unstoppable technology that became a vector for both issue-specific concerns and more general social and political anxieties. Beyond detailed challenges about nuclear safety and open-ended problems of nuclear wastes, wider issues presented themselves in intense forms. For both GM and nuclear power, these arguments reflected not simply 'technical' issues held to be legitimate by governments and scientists, but also wider social relations in which the respective technologies were embedded. In the absence of other meaningful spaces in which such debates could take place, GM became the occasion and the opportunity.

Lessons for nanotechnologies

So what implications can we draw from this account of the GM controversy for future approaches to nanotechnologies? First, when faced with new situations and technologies, regulators will usually turn to assessment frameworks developed for previous technologies and tied into existing debates. Given this tendency to 'fight the last war', there is a need for more textured, socially realistic analysis of the distinctive character of particular technologies, and greater recognition of the limitations of conventional models of risk assessment.

Second, it is important to be more realistic about the diverse roles of NGOs. The breadth and unfamiliarity of issues now being thrown up by new technologies mean that NGO responses are in continuing flux, and a richer account of the ways in which NGOs 'represent' opinion in wider society is needed.

Third, the GM case suggests that the deficit model of public scepticism or mistrust of science and technology is a fundamental obstacle for institutions charged with the regulation and assessment of new technologies. For nanotechnologies, there is a need to build in more complex and mature models of publics into 'upstream' policies and practices.

Fourth, GM demonstrates the ways in which new technologies often operate as nodal points around which wider public concerns condense. Such processes of condensation are inherently unpredictable. However, a richer understanding of the underlying dynamics of such processes – informed by recent thinking in the social sciences – could begin to provide some clues. In considering approaches to the social handling of nanotechnology and its potential manifestations in applied forms, care will need to be taken to 'design in' greater social resilience.

Finally, the GM experience highlights the degree to which scientific research is informed by imaginaries of the social role of technology. With GM, these tacit visions were never openly acknowledged or subject to public discussion. For nanotechnologies, a more open model of innovation is required, in which imaginaries are opened up to greater scrutiny and debate. But first, we need to understand what those imaginaries are, who holds them, and how they are influencing research and policy agendas. The next chapter describes our efforts to do this, by spending time in two nano laboratories and interviewing a range of prominent figures in nanotechnology.

3. Laboratories of imagination

Bruno Latour, a philosopher of science, once declared: 'Give me a laboratory and I will raise the world.'[36] By this he meant that implicit in laboratory practices and the organisation of science were a whole range of assumptions in which the relations between the inside (the technical) and the outside (the world) are never clear cut.

Understanding such processes has been the preoccupation of science and technology studies for the past 30 years. Stemming from anthropology, history and sociology, science and technology studies have challenged the self-description of science as apolitical and asocial, and have highlighted the contingent and situated nature of scientific knowledge.

Historical considerations of the societal implications of science and technology have often ignored the technical particularities of scientific practice. Indeed, the internal workings of science have traditionally been 'black-boxed'. The assumption is that there can be no 'bad science', only 'bad technology'. For thinkers such as Latour, this logic is faulty. Rather, science is deeply social and cultural. In order to uncover the societal dimensions of science and technology one must open the black box and examine how science is practised:

We will not try to analyse the final products, a computer, a nuclear plant, a cosmological theory, the shape of a double helix, a box of contraceptive pills, a model of the economy; instead we will follow scientists and engineers at the times and places where they plan a nuclear plant, undo a cosmological theory, modify the structure of a hormone for contraception, or disaggregate figures used in a new model of the economy. . . . Instead of black boxing the technical aspects of science and then looking for social influences and biases, we realise . . . how much simpler it is to be there before the box closes and becomes black.[37]

This means attending to how science is practised – in real time – and the institutional and cultural contexts in which it is situated.

The appliance of science

So welcome to the nano lab. It is in such places that the relationship between grand visions of nanotechnology and the practical business of research is worked out. In the second phase of our research, we spent a few weeks with nanoscientists at the Cambridge Nanoscience Centre and the Department of Material Science at Oxford University. During this time, we conducted interviews with nanoscientists and observed lab practices.[38] This was followed by a series of interviews with leading figures in the global development of nanotechnology.

Establishing the Cambridge Nanoscience Centre – and others like it in London, Bristol, Manchester and Newcastle – is an attempt to coordinate and centralise research in one building. It is in this building that futuristic images of nanoscience are being realised, with its collection of state-of-the-art clean rooms, open-plan offices and informal meeting spaces. The building itself speaks of 'new science'. It says 'gone are the days of the individual researcher tucked away in a biology department'. Instead, nanoscience is creative, interdisciplinary and built on teamwork.[39] The Centre reflects a range of expectations regarding the promise of nanotechnology – that it will lead to new applications and assist in maintaining the position of UK science in a globalised knowledge economy.

And yet, the first thing one notices at the Centre is the strategic decision to call the laboratory a nanoscience centre rather than a nanotechnology centre. Many of the researchers we spoke to stressed the importance of maintaining a clear distinction between nanoscience and nanotechnology, and suggested that their primary interest lay in the 'basic science' in which applications were not immediately evident.

The complex connections between basic nanoscience and applied nanotechnology are evident in the way the research at Cambridge is structured. The Cambridge Nanoscience Centre is the lead partner in a larger network – an Interdisciplinary Research Collaboration (IRC) in Nanotechnology, which involves Cambridge University, University College London and Bristol University. The original proposal for the IRC outlines the intention to undertake basic science that will underpin developments in nanotechnology:

> *The IRC proposed here is directed at the very core of nano-technology and as such will aim to provide an underpinning interdisciplinary activity with the general theme of fabrication and organisation of molecular structures. . . . The consortium will develop the basic tools to organise molecules at the hard/soft interface (the growth of 'soft' molecular structures off 'hard' substrates) by natural and other means, including self-assembly and soft lithography.*[40]

What is going on here? It is clear that nanotechnology is fuelled by a range of expectations and promises.[41] It is imagined that 'basic' nanoscience will provide the intellectual and conceptual bedrock that will enable the development of new applications. While many of the researchers we spoke to claimed to be interested primarily in the basic science of nanoscale research, there are also many hints around the laboratory of broader technological and economic drivers. For example, in the communal kitchen area, a picture of a Japanese train was pinned to the notice board, with a scribbled caption explaining the speed it was capable of achieving. Lying around the office space were many industry magazines and a well-thumbed copy of Michael Crichton's novel *Prey*, which outlines a world in which nanobots swarm out of control.

The IRC proposal document hints at potential tensions that exist between curiosity-driven research and commercial application, and outlines its model for managing technology transfer. It continues:

> This 'basic science' nucleus of activity will be kept focussed, and will be aimed at encouraging curiosity-driven research. However, this will not be at the expense of relevance to industry and to applications. . . . We identify at the outset two broad classes of 'end-user'. The electronics/communications industry will benefit from the development of electronics and photonics with molecular or polymeric materials. Applications in the biomedical area will result from advances in tissue engineering and biosensor technology.[42]

The laboratory then is a place where tensions between basic research and technological application are being renegotiated constantly. Expectations of what nanoscience can deliver meet the science itself. Our research identified five imaginaries that are influential within the lab and beyond:

○ nanotechnologies as an extension of the 'miniaturisation imperative'
○ nanotechnologies as 'control over the structure of matter'
○ nanotechnologies as a 'revolution'
○ nanotechnologies as a 'new science'
○ nanotechnologies as 'socially robust science'.

These imaginaries do not necessarily or directly influence what research is done, or determine what is established as 'basic science'. Nevertheless, the idea that social and policy responsibilities begin only at points of decision about application, and never before (in the 'pure' science), is called into question by the ways that such broad-brush imaginaries operate as a selection criteria for what is seen as a salient technical question to research. If we test this proposition intellectually by thinking of such putative influences only as deliberate *decisions*, then we are likely to miss crucial influences which occur beneath this threshold, as the influence of inadvertent, taken-for-granted, undeliberated commitments shaped by the routinised cultural elements of scientific laboratory worlds.

Basic scientific knowledge does accumulate through these socially imbued practices; but this does not mean that it has been biased by 'social interests', or rendered technically false, as is sometimes feared. This is a major finding of our research, and one that poses a challenge to advocates of the public engagement movement, who must now identify ways of opening up such imaginaries to scrutiny and accountability.

Nanotechnologies as an extension of the 'miniaturisation imperative'

Nanoscience and nanotechnology are typically cast as the natural extension of developments within scanning tunnelling microscopy (STM). The scanning tunnelling microscope was invented in 1981 by Gerd Binnig and Heinrich Rohrer at IBM's Zurich Lab and enables the visualisation of regions of high electron density, and hence the position of individual atoms. The technique employs a sharp probe (or tip) that is moved over the surface of the material under study. Coupled with the existing miniaturisation of IT components – particularly using lithographic techniques – STM technology is regarded as one of the direct forbears of nanotechnology because it enables researchers to 'see' atomic patterns and shapes, and because in imaging individual atoms, the scanning tunnelling microscope physically alter surfaces in ways that enable the movement of individual atoms.[43]

However, the lineage between nanotechnology and STM is not purely technical. The timing of the invention of the STM in the early 1980s is significant. This was shortly after the word nanotechnology was first coined by Norio Taniguchi in his 1974 paper[44] and shortly before K Eric Drexler published his well-known account of nanotechnology, *Engines of Creation*.[45] In the same period, the UK launched a National Initiative on Nanotechnology (NION) with strong backing from the UK X-ray and microscopy communities. In the 1980s there appears to have been a ferment of ideas for research at the nanoscale, in which developments in STM played a significant role.

The fact that STM was developed by IBM is also important. Not only did IBM become an early leader in nanoscience research – stamping a degree of ownership over the nanoscale by positioning 35 xenon atoms in the shape of the IBM logo – it also reinforced the imperative towards miniaturisation of electrical devices. Through this trajectory, nanotechnology therefore became firmly rooted in developments of micro-electronics and advances in data storage.

Nanotechnology is positioned as the 'natural' inheritor of the drive towards miniaturisation in electronic circuits and data storage. As such, nanotechnology is also guided by the economic and technological imperative to maintain Moore's Law – the prediction that the capacity of an integrated circuit doubles every 18 months. Moore's law operates both as a prediction of what *will* happen and an imperative within the semiconductor industry to maintain a particular rate of technological development. Phil Moriarty, a professor at Nottingham University, explained this to us:

> *Certainly in terms of the natural progression of technology, whether it's the buzz word of a paradigm shifting . . . [nanotechnology] is absolutely necessary. It's fundamental to maintain anything approaching Moore's Law, anything like that. . . . Because silicon is going to run out of steam, conventional technology is going to run out of steam.*[46]

Many of the scientists talked about the goal of making smaller electrical circuits and denser forms of data storage. Commonly cited examples such as the 'iPod nano' are seen to confirm a public appetite for ever smaller, faster and smarter electronic components. The perpetual miniaturisation of technology is cast as both inevitable and desirable. This researcher in Cambridge explained to us the link between his research and applications in this field:

The big potential application of much of the work I do is in some form of data storage. . . . The technological driver is simply to make those storage entities as small as possible because then you get more data per square inch.[47]

Yet although the researcher identified a technological driver for his work, he does not see himself as responsible for the task of following his research through to development and the market. This acknowledgement of the diversity of roles in innovation was a characteristic response of the nanoscientists. One researcher in Oxford explored this with us:

There are always potential applications. But I'm personally quite a strong believer you find applications when you do things. It's like when people were first making semiconductor diodes, everyone was, like, well it's a really interesting point, but what's it going to do? Then 10 years later everyone has got a solid state computer. Basically that's using exactly the same technology when people said what's the point of it. And so I'm quite a strong believer that if there is research being done, people will find ways to apply it.[48]

Nanotechnologies as 'control over the structure of matter'

Nanotechnology is also imagined as a form of 'control over the structure of matter'. Christine Peterson, of the Foresight Institute, described this to us:

I think what we're aiming at is the total control of the structure of matter. The layman's phrase they use is 'building atom by atom' which is not technically exact. I think it is more you are building molecule by molecule. But the meaning they are trying to get across is building with atomic precision. . . . So that is revolutionary in terms of what you can build, how cleanly you can do it and I think it will have revolutionary effects on human life.[49]

The suggestion that this can be achieved through replicators and universal assemblers, popularised by Eric Drexler, has been critiqued extensively.[50] But while many of Drexler's ideas have been rejected by mainstream science, his broader goal of 'control over the structure of matter' has emerged as a working vision for nanoscience. For many of our interviewees, this had crystallised into *the* project for nanotechnology. Nadrian Seeman, the DNA nanoscientist, explained:

> *Our real goal is control over the structure of matter in 3D. To be able to say . . . I want this thing to be in this place relative to other things. And I want it to perform in this fashion over time. . . . Understanding and control go together. Feynman once said, 'If I can understand it I can create it.' If it was good enough for Feynman it is good enough for me.[51]*

Though discussed in technical terms, this imaginary is deeply social, political and cultural. The ability to operate at the nanoscale – atom by atom – symbolises an expression of power. It represents the material world subordinated to human will with unprecedented degrees of precision and control. More pragmatically, the suggestion that *once* we have achieved control over the structure of matter *then* the radical possibilities of nanotechnology will be realised is an implicitly linear model, which omits a number of steps along the way.[52]

In our research at Cambridge and Oxford Universities we found a similar pattern – that despite the repudiation of Drexler, the drive towards accuracy, control and precision has become an assumed part of the purpose of nanoscale research. One researcher at Cambridge summarised the aims of her research as:

> *both precision and control. . . . Ideally you would like to reach a goal that science has worked towards ever since it was created . . . to enhance life itself and to have the precision to know what you need to target and target it immediately.[53]*

Though this programmatic vision was accepted by many of the researchers we spoke to, and internalised in their own work, we also found that many were willing to acknowledge the complexities involved in realising this vision. Another researcher at Cambridge described how a priori assumptions of control and accuracy are moderated by actual research practice:

> *Most of it is looking at what biology has done and saying 'right, I'll have that'. So you take it out of the biological context. And sometimes they just do something that you don't expect at all. We don't have the mechanisms to predict. We can't say 'right I'll take this' and it will do this. We're just not advanced enough in our biology. It's a horribly complex system. It is possible to achieve control, but only by trial and error.*[54]

Understanding such errors, by reviewing and correcting the assumptions that were shaping them, is the bread-and-butter of productive scientific research. It is interesting to note, however, how at this detailed level things are more accidental than the idealised model of 'pure science' suggests.

Nanotechnologies as a 'revolution'

The next dominant imaginary is of the potential social and economic impact of nanotechnology. An early document by the US National Science and Technology Council in 2000 was entitled *National Nanotechnology Initiative: Leading to the next industrial revolution*. It stated that:

> *The emerging fields of nanoscience and nanoengineering – the ability to precisely move matter – are leading to unprecedented understanding and control over the fundamental building blocks of all physical things. These developments are likely to change the way almost everything – from vaccines to computers to automobile tires to objects not yet imagined – is designed and made.*[55]

In this we see a combination of visions of control and precision and of the technological imperative to perpetuate innovation. It is predicted that nanotechnology will change 'almost everything'. The message is clear: the impact of small technology will be big.

This revolutionary potential of nanotechnology is also evident in the ways in which potential applications are discussed. Mike Roco and William Bainbridge, two of the main architects of US nano policy, suggest a list of almost limitless possibilities:

○ *Fast, broadband interfaces directly between the human brain and machines will transform work in factories, control automobiles, ensure military superiority, and enable new sports, art forms and modes of social interaction.*

○ *Comfortable, wearable sensors and computers will enhance every person's awareness of his or her health, environment, chemical pollutants, potential hazards and information of interest about local businesses and the like.*

○ *The human body will be more durable, healthier, more energetic, easier to repair, and more resistant to stress, biological threats and aging processes.*

○ *National security will be greatly strengthened by lightweight, information-rich fighting systems, uninhabited combat vehicles, adaptable smart materials, invulnerable data networks, superior intelligence-gathering systems, and effective defences against biological, chemical, radiological and nuclear attacks.*

○ *The vast promise of outer space will finally be realised by means of efficient launch vehicles, robotic construction of extraterrestrial bases and profitable exploitation of the resources of the Moon and Mars.*

○ *Agriculture and the food industry will greatly increase yields and reduce spoilage through networks of cheap, smart sensors that constantly monitor the condition and needs of plants, animals and farm products.*

○   *Transportation will be safe, cheap and fast, due to*
    ubiquitous real-time information systems, extremely
    high-efficiency vehicle designs, and the use of synthetic
    materials and machines fabricated for optimum
    performance.[56]

Even in the more conservative language of a 2002 report from the UK
government, the sheer promise of nanotechnology is palpable:

> *Few industries will escape the influence of nanotechnology.*
> *Faster computers, advanced pharmaceuticals, controlled drug*
> *delivery, biocompatible materials, nerve and tissue repair,*
> *surface coatings, better skin care and protection, catalysts,*
> *sensors, telecommunications, magnetic materials and devices –*
> *these are just some areas where nanotechnology will have a*
> *major impact.*[57]

Such lists of the possible applications of nanotechnology appear so
broad as to construct nanotechnology as a cure for all human ills, and
as the sustainer of economic growth and prosperity. There is also a
strong determinism in many of these predictions – Roco and
Bainbridge claim for example that 'comfortable wearable sensors and
computers *will* enhance every person's awareness'. Hypothetical
developments are presented as imminent and inevitable.

Another concern of many governments is the perceived need to
'keep up' in the global race to develop nanotechnologies. Although
this fear of being left behind is endemic to many scientific fields, the
predicted revolutionary impact of nanotechnology intensifies these
fears. In the UK, it is often suggested that after taking an early lead in
nanotechnology research – through the NION in 1986 and a LINK
programme in 1988 – the UK fell behind its competitors. In 2001 and
2002, the Department of Trade and Industry (DTI) sent missions to
Germany and USA, both of which reported that the UK had lost any
first mover advantage that it gained through those early initiatives,
and was now lagging behind:

In 1986, the UK was on the threshold of opportunity; in 2001 we are on the threshold of a major threat. There is still time to address this, however, but we need to start immediately, build momentum quickly and coordinate our activities across academe, industry and government. Given the impact that nanotechnology will have on employment, wealth and technological capabilities in the UK, the 'status quo' is not an acceptable option.[58]

In response, the DTI published a report in 2002, *New Dimensions for Manufacturing: A UK strategy for nanotechnology* (commonly known as the Taylor Report). This was critical of the lack of a coordinated nanotechnology strategy:

[The] national and international perception of the UK's research in nanotechnology is coloured by its fragmented and uncoordinated nature. It is seen to be dominated by a number of internationally recognised individuals rather than there being world-leading UK centres. The UK is not recognised as having a critical mass of world-class activity, but is seen as having a thinly spread network of leading players.[59]

In response to predictions of a nanotechnology revolution, many of the scientists we interviewed were more cautious and pragmatic. One researcher at Oxford expressed a degree of ambivalence in relation to government strategies for supporting nanotechnology:

You could always ask the question: 'What would happen if it wasn't labelled nanotechnology, or what would happen if there wasn't a large pot of money ring fenced for nanotechnology?' Would it happen anyway? If they are going to be good products would they get out there anyway?[60]

Another researcher at Cambridge questioned the expectations that his science should be used as an engine for innovation:

I've no interest myself in applying the science which comes out. I'm interested to see what happens but I'm not interested in being entrepreneurial with the material. . . . The goal as I see it is to do research and to get the results into the literature and see what other people do with it. . . . I mean Gordon Brown and others are perhaps trying to redefine my job to be something different and I've a lot of sympathy with that but I'm not convinced I'm necessarily the right person to be doing that.[61]

Nanotechnologies as a 'new science'

A related vision emphasises nanotechnology as a new kind of interdisciplinary science. This refers not only to a confluence of traditional disciplinary traditions, but also to a new model of research that is intellectually open to new ideas and structurally open to new partnerships with the corporate sector. There is a common assumption that nanoscience is *necessarily* interdisciplinary and that this will bring with it an increase in the rate of technology transfer between academia and business. For example, the Taylor Report says:

The top-down and bottom-up nature of nanotechnology underlines its multidisciplinary nature. Nanoscience and nanotechnology depend on contributions from, among others, chemistry, physics, the life sciences and many engineering disciplines. Thus the subject inevitably crosses the boundaries of many different departments in traditional universities and research institutes.[62]

Similarly, Renzo Tomellini, head of the Nanoscience and Nano-technology Unit at the European Commission, argues:

Interdisciplinarity is becoming more and more important. At the same time, linearity is in general no longer a valid model for production. Complexity has to be managed, and I think that we should be able to work through an enlarged partnership.[63]

One researcher at Cambridge suggested that working at the nanoscale necessitated collaborations across disciplinary lines:

> *You cannot realise certain things by just studying one subject. You have to either rely on other people's expertise, or you try to understand everything. Of course nobody can do that. So basically what you do is you borrow experience and expertise from other people, from other disciplines. My role is to combine them together to do a new science. We call that fusion science.*[64]

This idea of nanotechnology as interdisciplinary is also seen as a model for how modern science should operate. Mark Welland, director of the Cambridge Nanoscience Centre, explained:

> *The unique thing that we tried to do [at Cambridge] is to be interdisciplinary. . . . We wanted to make sure that people from a number of university departments actually worked together in one place. . . . So we genuinely felt there was a need to have people working together, sat next to each other. . . .*
>
> *The research councils when they judge us will measure us by how many publications are by somebody from physics, somebody from chemistry, somebody from biology mixed together. How many people have started off in one discipline and ended up in a job in another one?*[65]

Nanotechnologies as 'socially robust science'

A final imaginary concerns the importance of public dialogue and social science in helping to shape the development of the technology. For example the US National Nanotechnology Initiative report, *Leading to the Next Industrial Revolution*, mentions the need for more social research:

> *The impact nanotechnology has on society from legal, ethical, social, economic, and workforce preparation perspectives will be studied. The research will help us identify potential problems*

and teach us how to intervene efficiently in the future on measures that may need to be taken.[66]

In the UK, the Royal Society/RAE report called for 'upstream' public dialogue in setting the agenda for nanotechnology, and noted that:

Our research into public attitudes highlighted questions around the governance of nanotechnologies as an appropriate area for early public dialogue.[67]

In both cases, nanotechnology is being cast as an opportunity to build a new type of socially robust science. As with the other programmatic visions, we found there to be a partial degree of internalisation within laboratory practice. Many of the researchers interviewed were not aware of the Royal Society/RAE report, but did recognise their social responsibilities as publicly funded researchers. A researcher at Oxford explained:

We are all living in the real world and we do rely on public money to do our work and so, because of that, public perception of our work is very important. There have been a few examples of other technologies recently which have probably not done as well as they should have done in terms of public perception. . . . I am very hopeful that nanotechnology won't go down that same route. . . . Because if public perception turns against a particular form of research or technology then it makes it much more difficult for a government to justify putting resources into it.[68]

While some researchers recognised instrumental reasons for ensuring public support for nanotechnology, others had difficulties in conceptualising public concerns. One researcher at Oxford argued 'if nanotechnology is going to be a bad word then let's call it something else'.[69] She felt that nanotechnology was a convenient label for attracting funding, but could be jettisoned in the event of adverse public reactions.

Our time spent in the lab confirmed the idea that visions and imaginaries play an important role in nanotechnology research. But while funding is often based – at least in part – on these promises and expectations, the relationship between programmatic visions and the everyday practice of nanoscience is complicated.

Many researchers acknowledge the wider motivations for nanoscale research and try to negotiate a space within these for their own research interests. This is common and to be expected. More interesting are the gaps and tensions between structural visions of the economic and social promise of nanoscience, and researchers' own experiences. Many researchers lacked a sense of agency or felt ill-equipped to interpret the interconnections between science as they practise it in their specialist worlds, and 'science' as an object of policy and commercial design, control and expectation. This suggests that scientists need more opportunities and encouragement to reflect on the societal dimensions of their work. One way of doing this is through direct engagement with wider publics, and it is to this challenge that we now turn.

4. Nanotechnologies in focus

In a north London suburban living room, a group of women in their 30s and 40s sat discussing what life would be like in a world of radical nanotechnologies. The conversation twisted and turned as different dimensions of the technology were articulated. Would our skin still wrinkle? Could we stay looking 30 years old for our entire lives? Would everything be '100 per cent perfect and plastic'?

Much of the conversation centred on the latest anti-wrinkling cream, newly marketed by L'Oréal as containing 'nanosomes' for increased effectiveness. After a week of reading about nanotechnology on the internet, the women were concerned. Did anyone know about the long-term effects of nanoparticles? What tests had been carried out? By whom? And given all these uncertainties, why aren't new cosmetics regulated in the same way as new drugs or foods?

This was just one of a series of ten focus group discussions that we ran in the third phase of the Lancaster–Demos project. Building on our analysis of the role of imaginaries in shaping scientists' expectations, the purpose of this phase was to develop a deeper insight into the sorts of issues likely to shape public attitudes and concerns. This was not an easy task. How do you research a topic about which most people have little or no opinion? How do you anticipate future public opinion?

Nano publics

In the past few years, a number of studies have examined public attitudes towards nanotechnologies. These include an item on the 2002 Eurobarometer survey,[70] a UK survey and set of focus groups run on behalf of the Royal Society and Royal Academy of Engineering in 2004,[71] and two US surveys conducted by the Woodrow Wilson Center in Washington, one with a large sample, the other small but in-depth.[72]

By and large the above surveys have followed a familiar form, starting with questions designed to measure public knowledge and awareness of nanotechnologies, followed by perceptions of risks and benefits, and finally leading to questions of trust in industry and government regulation. The results suggest that most Americans and Europeans are unfamiliar with nanotechnology,[73] that most people anticipate benefits to outweigh any risks, that the most negative aspects of nanotechnologies are perceived to be 'loss of privacy' and self-replicating organisms,[74] that there is high demand for regulation and public information,[75] and that on both sides of the Atlantic there is suspicion of the motives of industry and little trust in government.[76]

Yet while this survey data offers an indication of how people respond to the term nanotechnology and its potential trajectories, it nevertheless provides only a limited insight into the underlying dynamics likely to structure public responses. There are several reasons for this.

First, nanotechnology is an open-ended and disputed term. Defined simply by scale, it transcends disciplinary domains and sectors of application. Second, many nanotechnologies remain at an early or pre-market stage of development. Although nanoparticles are being used in a growing range of consumer products,[77] much still exists only in the 'promise' of nanotechnologies to radically restructure future economies and society. Third, given that most people are unfamiliar with the term nanotechnology, and so presumably do not have pre-existing attitudes, it is not clear what these survey questions are measuring. And finally, surveys tend to frame the public dimensions of concern within a broadly conceived 'risk' and 'benefit' rubric. While such framings fit within official regulatory and risk assessment vocabularies, it remains an open question whether they reflect public sentiment. Indeed, in the related domain of GM foods and animals, a clear research finding was the inadequacy of such official framings in capturing what people actually feel to be 'at stake' in the application of biotechnologies.[78] Public attitude surveys tend to compel respondents to adopt 'attitudes' towards a technology while ignoring the factors underpinning the formation of such attitudes.

These various constraints and imponderables pose difficulties for social research aimed at understanding future public responses, at this early stage of their emergence. As a result, our project used a three-stage focus group methodology that we felt could generate more insight in these circumstances.[79]

The methodology

The purpose of the focus groups was to encourage discussion of potential issues arising for nanotechnology within a framework set by participants rather than imposed by official regulatory and risk-assessment vocabularies.

The sample consisted of five groups, each of which met twice, with a gap of one week between the sessions. Participants were recruited on the basis of their existing participation in local community or political issues, but with no prior involvement or exposure to nanotechnology. They included a group of **professional men** (doctors, architects, civil servants etc); a group of **professional women** (mostly employed as middle managers in business); a mixed group with demonstrable **political** interests; a group of **mothers** with children of school age; and a mixed group with an interest in **technology**. The groups were conducted in Manchester and London.

The groups were designed to allow space for participants to develop a collective imagination on a topic that was likely to be seen as unfamiliar and esoteric. For this reason, the groups were run on two consecutive sessions.

The first session began with a general discussion of new and emerging technologies, how they were affecting everyday life, in what ways they were giving rise to 'social' questions, and what people imagined to be the key issues for the future. Halfway through the session, the concept of nanotechnology was introduced. Participants were next presented with some everyday consumer products that had been fabricated using nanotechnology, including a golf ball, a tub of anti-wrinkle cream, and a stain-resistant shirt. Using a set of concept boards as a stimulus, people discussed three different visions of nanotechnology: a mainstream view, focused on incremental developments and economic benefits; a radical utopian perspective, which emphasised more disruptive implications for society; and a sceptical outlook, which focused on potential risks and negative social implications.

At the end of the first session, participants were asked to spend the week before the next session exploring the issues with friends and colleagues, consulting websites and keeping a journal for any reflections arising.

The second session explored how participants' perceptions and responses had evolved through their own discussions and research, followed by a discussion of particular social and ethical dilemmas. A second set of concept boards was designed to stimulate discussion in three areas: privacy and security, therapy and bodily enhancement, and the relationship between scientific progress and 'meddling with nature'. The session finished with a discussion of wider governance implications.

A third phase involved the selection of 12 people from across the five focus groups who were willing to take part in the day-long meeting with nanoscientists at the Natural History Museum (discussed in chapter 5).

Enthusiasm and ambivalence

Many participants had a sense of enthusiasm about technology and the ways it had improved the quality of their lives. It was felt that recent developments, especially in information technologies, had become so thoroughly integrated into everyday life that people found it difficult to imagine doing without email or their mobile phone:

M   The change that it's brought to every aspect of our lives, whether it's communications, phones, faxes or whatever, whether it's the cars we drive, the homes we live in, everything has been hugely impacted by technology, everything. The way we live our lives, the way our children live their lives, it cuts through every single aspect of your life. From the business aspects of your life through the home side of your life, entertainment, communication, travel, transport. Every single thing that you do or touch or have any involvement with has dramatically changed with technology. I would say all to the good, there are very few negative aspects of technology I can think of off the top of my head.

Professional man[80]

For the younger participants, technology had become a matter of lifestyle choice, with the pressure to 'keep up' now an internalised dimension of social life:

F   Technology is like an accessory almost isn't it, it's like the new Gucci handbag . . .

F   It's quite aspirational, isn't it . . .

F   There's always a pressure, isn't there?

M   There's definitely fashion within technology, there's no doubt about it. Magazines are just full of cool gadgets that look cool and have charts and ratings for which are the

best and things like that. Snap on covers so you can brand
it to yourself, make things look more personalised . . .

M   You're judged, aren't you?

Technology group

However, as the dialogue continued, participants started to discuss
ways in which technological life was double edged. The downsides of
technology were seen to include a loss of community, the decline in
courtesy and social relationships, invasion of privacy, erosion of
family and work boundaries, and information overload. An
interesting exchange occurred in one of the groups, as to whether a
more technological society meant we were any happier:

M   Are we any happier than we were before this stuff
happened? And the point of the telex going to the fax to
the email, the communication's still happening but really
it's just happening faster. . . . Whether that's necessarily a
good thing I'm not sure. . . . I mean I think it is for the
good but I don't know whether I'm any happier because
of it. It is for the good because we get better
communications, we get better service from our domestic
machines and we get better quality television. Whether
I'm happier now watching my choice of 50 odd channels
on cable than I was watching *Dixon of Dock Green* on
black and white is debatable.

Professional man

The discussion developed a different texture, too, when people were
asked to imagine the social impacts of technology in the future. The
anticipated pace, scope and intensity of technological change, and its
associated disruptive impacts on social life, were the source of
considerable concern. This was compounded by the lack of power
people felt they had to shape the direction of technological innovation. There was a sense that the ownership and control of technology were likely to be further consolidated into large and
unaccountable actors, outside the reach of citizens and national

governments. Such sentiments were expressed in discussions on genetic technologies:

> F   Well, things like genetics. . . . It would be interesting to see how it plays out. But I don't feel that I have control over [or] any input into how that happens, you know like cloning or genetic modification . . . it's rushing very quickly ahead. I don't ever feel like that's been an election issue or been in someone's manifesto. These sorts of things I think are going to be really big questions for humanity and I think that they're not really on [anyone's] agenda; but I don't feel that there is any way that we can – that I can – express my opinion.
>
> *Professional woman*

This sense of discomfort was aggravated by perceived inadequacies in the ability of the political system to address ethical, social and health implications of new technologies in advance of their application:

> F   What you're saying is we haven't had a say again. In that these things are just coming through and . . .
>
> F   They don't feel the need, no.
>
> F   But also the speed with which things are going forward as well, like I was trying to say before, I don't know, there are a lot of well-publicised questions around genetic modification which . . . I don't feel have been addressed, ethical questions haven't been addressed really or publicly. . . . I'm a little bit wary about jumping into, rushing forward with another new technology where I feel that the old questions haven't even been addressed. . .
>
> *Professional woman*

Perceptions of nanotechnology

Unsurprisingly there was little knowledge or familiarity with nano-technology. When pressed, people tended to define it as something that was scientific, clever, small, possibly medical, futuristic and associated with science fiction:

Int So when I say [nanotechnology] what comes to mind?
M   Alien.
F   Very little understanding of it.
F   Very scientific.
M   Well I do think quantum theory and strange, strange effects at that kind of level.
M   You just think it's so futuristic that it wouldn't be in our lifetime but then you think the way things are going so quickly.
M   Bewildering really.

Technology group

F   I know the idea that nanotechnology is really small technology and occasionally I'll read something in the *Guardian* or wherever about 'it's amazing, these guys have written their names in atoms on something' and you're like, wow, that's cool. And you have this very nebulous notion that this is really clever, you're told there are all these possibilities that are waiting to be unlocked in nanotechnology. But I actually have no idea, you know, what they're really doing and what these possibilities are. I just have this very vague notion that it's very clever and it could be really important.

Political group

It was fascinating to watch how participants developed their collective imagination of nanotechnology, and the factors that shaped their evolving opinions. Typically, the progression ran as follows: from a state of initial ignorance, to surprise at how much research funding was being invested by governments and industry, to enthusiasm as to the radical potential for social good (particularly in the medical domain), to unease and anxiety that nanotechnology might lead to unanticipated and disruptive problems. Participants also struggled to envisage the scale of nanotechnologies:

M   What I'm struggling to visualise is what they're [nanotechnologies] actually producing and what they're doing, it just seems incredible that something so small, you know, what is it replacing . . . ?

M   It is hard to grasp the concept of what sometimes is actually going to be. In other areas it's a bit vague about yeah somebody works some magic somewhere but how it'll integrate into the way you live is a different thing.

M   Just exactly that. It's just so difficult to grasp.

Technology group

The groups then explored the three visions of nanotechnology – mainstream, utopian and sceptical – depicted on a series of concept boards. Responses to the mainstream vision ranged from genuine surprise as to the extent of investment in nanotechnology research, to scepticism as to whether such investment would bring any real benefits. Much of the discussion centred on how commercial considerations are likely to drive the technology:

F   I am a bit cynical about it because my impression is that with these sorts of technologies there's so many great uses to help people you know, but it seems to me that those uses don't generally get through. The way that these technologies are applied are decided by the people who have the money to put behind them, and those are the corporations or whoever stands to make a profit. . . . [You] don't generally see technologies applied in a more humanistic or socially beneficial way.

Professional woman

Interestingly, it was the utopian vision of nanotechnology that generated the most negative responses. Its predictions of radical improvements in human capacities were seen as dangerous and hubristic by all the discussion groups:

M   That really is quite a frightening scenario that when you read through that. . . . So this wonderful nanotechnology is going to be a cure-all for all human ills, it's going to make us all super brilliant and clever and work that much better, our transport's going to be far better, even though the fact that nobody will be dying of old age, nobody will be dying of any illnesses, means we won't be able to move on this planet. . . . OK, if it's used to treat cancers and stuff like that but we're getting into this Brave New World scenario here where everyone lives forever and everybody has everything, everybody can do everything. . . . It's a very, very frightening scenario . . .

Professional man

F   It's like nanotechnology's the new God.

Technology group

While superficially appealing to some, these technological visions were seen to raise substantial moral and social issues, not least around the ability of governments and industry to exercise sufficiently robust forms of control and oversight:

F   It's amazing.

F   I find it quite daunting actually, I find it a bit scary.

F   This is the vision of the robotic environment with everything controlled for you and everything 100 per cent perfect and plastic.

F   It's like even the food. . . . You buy a piece of fruit, it's healthy, after a period of time it wrinkles, you throw it away or whatever and that is a natural process and I think in some ways it's kind of fiddling with that natural process.

Mothers

A common response was to appeal for such innovations to 'slow down' and not move ahead of wider public values. Several people referred to the way government and business mishandled GM foods and crops as an argument for a more cautious approach. GM was seen as a case where exaggerated promises had been made, potential risks discounted, and public concern ignored. Set in this context, it was seen as rash to rush ahead in an arena of new technological advance.

The evolution of concern

During the week between the two sessions, participants engaged in their own research on nanotechnology, through the internet or discussions with family and friends. At the start of the second sessions, the questions or concerns raised by this research were explored with the help of three more concept boards on the themes of 'privacy', 'human enhancement' and 'meddling with nature'.

For many participants, the greatest area of anxiety was in relation to nanoparticles entering and harming the body, either through cosmetics or foods. The invisibility of nanoparticles exacerbated this concern:

F   The face cream which has got very small nanoparticles in it . . . if I rub that on my skin, there's things going into my skin I'm not aware of. No one knows exactly what that's going to do and it might have long-term effects. Any little bit of dirt, like something that shouldn't be in there, pops into the cell, messes with the actual sequence of what that cell does and you know, that's so scary.

F   Yeah because it can happen without you realising whereas before if things were going to invade your body, you would see it happening.

M   It's the invisible threat.

Technology group

A visceral example of this dynamic was explored in the mothers' group. In the initial session, these women were largely enthusiastic at the prospect of consumer benefits from nanotechnology, particularly in ameliorating signs of ageing. Now, when confronted by uncertainty as to the toxicological effects of nanoparticles, they were more doubtful:

> F   Since last week I've completely changed my approach to these creams. When you said it had those nanosomes I thought, 'oh great, fantastic, I'd use it' – I wouldn't touch it now with a barge pole if you paid me money to put that stuff on my face. It's so frightening.
>
> F   I think we're very trusting as buyers in the market, we're very trusting of the products we're given. We're suddenly having to become very sceptical because things come out afterwards.
>
> F   Well you sort of assume it's always been tested.
>
> F   Yes.
>
> F   Clearly things like cosmetics don't have the controls that the drugs do.
>
> F   But surely wouldn't they be better to say, right, we don't know enough and until we know enough or we've changed our regulations then we don't let it go on the market.
>
> F   There's too much money in it I think.
>
> *Mothers*

The potential toxicity of nanoparticles was seen as symptomatic of the wider phenomena of advanced technology proceeding in the face of unanticipated risks. GM foods, MRSA and mad cow disease were mentioned as other examples. Nanotechnology was seen as a worrying extension of this dynamic, particularly because of its perceived ability to transform society and nature:

> F   I mean it's exactly what somebody over here said before, we're turning into robots. That is exactly what it sounds like.
>
> F   This is the threat.
>
> F   When it comes directly to human beings and trying to make them, it's like trying to make a perfect race again.

F   We just don't know the long-term effects do we? That's the problem.

F   So basically our generation's going to be the one that they test this all out on. If it all goes horribly wrong, we'll be the guinea pigs.

Mothers

M   It'll get out of the cage I'm sure and evolve through various biostrains and mechanisms and it will be adapted, possibly. There are cases with GM super weeds now.

Professional man

Another concern was whether nanotechnologies would enable government and business to take new forms of control over everyday life. There appeared to be a fine line between technologies that would enable choice and autonomy and those that would control and limit opportunity:

M   I think the worrying thing for me . . . is that it's almost as though we lose control of what's going on because the technology itself is capable of replicating and you know pretty much making its own decisions.

M   I think that is a big problem. It's like the thing you were saying with creativity as well. If the human controls the technology that's fine; as soon as it becomes the technology making all the decisions then that's when you have a problem because humans are completely different from a computer.

M   There are some scary dark futures where you have strains of children who are and are not enhanced in some way, and that's a really dodgy thing.

M   Do you have your kids injected at birth to enhance the way their muscles grow and things?

Technology group

An equally potent dynamic was the potential for nanotechnologies to be used for purposes not imagined by their original developers, especially in the new security environment. Several participants saw nanotechnology potentially being used by terrorists:

> M   The more I think of the dangers, the more evil applications I can think of using nanotechnology.
>
> M   Well I just find it quite frightening really. I think it's quite disturbing. The potential to harm seems to me to be greater than the potential for good if it gets into the wrong hands.
>
> *Technology group*

A number of people argued that nanotechnologies would increase social inequalities and concentrate power in the hands of large corporations:

> M   And the other feeling I was left with was it was almost like a nano race to be the first to do it – because the impression I got was that whoever really is the first to do it well is going to pretty much monopolise everything.
>
> M   Yeah it makes the rich richer and the poor poorer.
>
> M   The gap just gets bigger.
>
> M   I agree, but I don't necessarily think everyone's going to benefit from it.
>
> M   Oh no certainly not everyone. Only the very rich few.
>
> *Technology group*

All of these factors contributed to the perceived difficulty of establishing robust and effective systems of governance and regulation. It was generally seen as unrealistic to advocate a slower, more cautious approach to nanotechnologies. Some suggested that an overly precautionary approach could harm the UK's economy and lead to outward investment. Others observed that much innovation is transnational and increasingly beyond the control of individual governments. It was widely felt that the pressure for commercial return would lead to corners being cut:

> F The whole thing we've been talking about is that these things happen so quickly, why can't we slow it down? Is it going to matter that much if it is slowed down?
>
> F But say this country does that and slows it down then you're gonna go abroad . . . yeah, and it's gonna come back into this country anyway.
>
> *Professional woman*

Reflecting on the groups

Our research presents a picture of emergent public opinion which differs to some extent from the existing literature on public attitudes towards nanotechnology. It highlights a latent ambivalence towards nanotechnologies, and suggests that there might be some public unease about its potential implications. What is perhaps most interesting is that this ambivalence did not diminish through greater knowledge and awareness. Instead, through exposure to the multiple ways in which the debate was being characterised, and through debate and deliberation, our participants moved towards a more sceptical view as to the ability of government and industry to represent the public interest.

At the end of the second session, we asked participants to express their overall feelings towards nanotechnology and the likelihood of future controversy. The most common response was to feel nervous, apprehensive and unsure:

> Int How controversial do you think it's going to be?
>
> M Far more than genetic modification.
>
> M It's going to be more. And what are the fault lines through which it's going to become politically controversial?
>
> M The medical, the human biological angles as well as the food chain.

M   I would have thought in the present climate particularly terrorism. It must be an absolute godsend to the terrorists, this sort of technology.

Professional men

Although further research is required to corroborate the reliability of these findings across more diverse social groups, our research suggests that there is considerable – if latent – potential for controversy around nanotechnologies. It points to the density of issues – moral, social, political, as well as technical – posed by nanotechnologies and of the fundamental challenges for governance. And it suggests that the public can differentiate these issues, and deliberate their social meanings in more complex terms than simply as 'risks' and 'benefits'. These dynamics were explored further in the final stage of the project.

5. A meeting of minds

Conversation is, among other things, a mind-reading game and a puzzle. We constantly have to guess why others say what they do. We can never be sure when words will dance with each other, opinions caress, imaginations undress, topics open. But we can become more agile if we wish.

Theodore Zeldin[81]

Feeling suitably empowered by their deliberations, 12 of our focus group participants agreed to come to the Natural History Museum to meet some nanoscientists. The day was divided into two halves: a series of small discussion groups, followed by a larger plenary. The filmmaker Hugh Hartford recorded events and the artist Tim Caswell captured the flow of the discussion on a giant mural.[82]

The tone of the day's conversation was intelligent, open and realistic. Occasionally, one or two of the scientists fell into familiar arguments about public ignorance or disinterest. But these were quickly contested by the public participants:

Sci   I think we've spoken quite a lot about things that
scientists can do for the public. But it seems to be quite a
one-way process in that scientists have to make science a
bit more sexy; we have to do the public engagement. But
you can only take a horse to water so to speak and I get

the feeling that the public are only interested in science when it affects them. People don't actually care about science; it's a bit boring. And I think maybe if the public were a bit more willing and if they showed more interest and if it wasn't just about them and how it's going to affect their lives, it would be an easier process.

Pub I think you're wrong. I think they do care and I think they want to be involved but they're not allowed to be involved at the moment because we haven't got the facility to do that.

Pub We don't know what's going on, so how can we have a voice if we don't know what's going on?[83]

But such moments of real disagreement and disjuncture between the two groups were rare. Instead, a common set of understandings – even at times, a consensual language – emerged over the course of the afternoon, as members of the public developed a better sense of life in the laboratory, and as the scientists grew to appreciate the legitimacy of public concern. Reflecting on the event afterwards, Richard Jones, professor of physics at Sheffield University and the most senior scientist present, said:

I think what's important is not the narrow issue of 'Do you do this piece of science and don't you do this piece of science?' Rather it's 'What kind of world do you want to live in?' The things that worried the people in my focus group were the things that worried me. I am uncertain about how lots of this stuff will turn out. I have a positive view of how I would like it to turn out but there are people who have opinions about how it ought to turn out that I really don't like at all. It's quite reassuring to think that I am not alone in worrying about the things I worry about.

The methodology

The workshop was held on 5 November 2005 at the Natural History Museum. It was designed to provide a forum in which members of the public, who had previously been involved in the focus groups, could interact with a similar number of nanoscientists. The workshop comprised 12 members of the public and 12 scientists, six facilitators, a film crew and a visual artist who recorded the conversation in the form of a giant mural. Members of the public were selected to ensure an even spread across the five original focus groups. The scientists were invited as individuals or nominated by their department to attend. Invitations were targeted at scientists at an early stage in their careers.

The day began with a brief introduction and ice-breaker. In small discussion groups, members of the public and scientists questioned each other about what is at stake in the development of nanotechnology and its potential social implications. The workshop then reconvened as a plenary in which more general questions of governance, regulation and responsibility were discussed. The event concluded with a reflection by Brian Wynne, who attempted to weave together the multiple threads of conversation.

The workshop was both a research exercise and an opportunity for some members of the public to pursue their newly developed interest in nanotechnologies through exchanges with working scientists. As such, it was an experiment in public engagement, although its intention was never to produce a formal set of conclusions authored by the participants. As a direct channel of engagement, the workshop provided members of the public with the confidence and opportunity to talk directly to scientists.

As noted earlier, part of the challenge in exploring the social, political and ethical dimensions of nanotechnology is the 'upstream', underdetermined nature of nanotechnology. But despite these difficulties, this is a crucial stage at which to undertake such deliberations if they are to have real purchase on the development of the technology.[84]

Risk and regulation

The discussion quickly turned to questions of risk, control and regulation. As with the focus groups, there was concern about the potential risk of existing products using nanoparticles – particularly in cosmetics and sunscreens. The feeling was that these products had entered the market with insufficient public scrutiny or regulation. A member of the first focus group explained:

> Pub   Our group very much felt that in some ways it's not acceptable to a consumer to put a face cream out there and not have it explained that there could be nanoparticles in that.

The response to this was a demand for better regulatory control. Members of the public suggested that there should be new regulation for nanoparticles:

> Pub   That was one of the things that shocked us in that there wasn't any regulation in terms of the cosmetics as there was in terms of the medical use. I think we couldn't get our heads around that at all. That you could have it used in cosmetics and not have any regulation behind it at all.

Some scientists suggested that existing regulatory mechanisms were sufficient:

> Sci   From my understanding, because nanotechnology in many ways has grown out of existing scientific research ... regulation has sort of worked in the same way. ... People might now be kind of considering the nature of those regulations, maybe revising them. But it's not that

we're saying we've reached a new chapter of science and we have to sit down and draft a completely new set of regulations.

Regulation is a relatively easy topic; it allows scientists and publics to discuss technology within assumed parameters. Participants were also able to imagine their own agency in decision-making, because of the presumed democratic accountability of regulatory processes. Regulation represented one way in which ordinary people might be able to take some control, whether through labelling or other mechanisms:

Pub   The fact that it's got to be on a product label gives you a choice; it gives you some power.

But there was also a recognition of the disconnect between the choices that we make as individual consumers, and bigger challenges that require new forms of governance:

Pub   One thing that we touched upon and was certainly of a lot of interest to me was the fundamental kind of global–local tension in terms of . . . how the scientific method works, in terms of the way research is geared towards stripping away all that's irrelevant to look at one tiny piece of causality . . . but the things we're creating, the applications of it are a global thing on a global scale.

Who is in control?

Demands for regulation, labelling and legislation led to discussion of who would regulate and whose voices would be influential in informing regulatory procedures:

Sci   One of the things I was intrigued by is who influences the processes before we get to the point of the regulation. Because by that time you're already down a very, very long time scale of previous innovations.

The question of *who* influences the process of developing regulation reflects the importance of the 'upstream' framing of science and technology. Discussions of the funding of nanotechnology research led participants to questions of purpose, need and equity, over and above the particular dilemmas of specific nanotechnologies. The discussion focused the role of public voices in this process and notions of responsibility. One factor that was discussed was the growth of private sector influence in universities, and the associated move towards more closed forms of proprietary knowledge. Two of the scientists revealed their concerns:

Sci   More and more companies are closing down their R&D
        facilities and outsourcing it to the universities. And I
        think it is something we have to resist. As public
        employees in universities, it is our obligation not to be
        drawn down those lines.

Sci   I feel very disheartened when I have to sit in a
        seminar room with friends presenting their subject
        and one of them gives out non-disclosure agreements
        and I have to sign it else I cannot stay in the seminar. . . .
        And I think that shouldn't happen and it happens
        increasingly.

Agency and responsibility

Many participants reflected on their sense of powerlessness when confronted by innovation trajectories. Antipathy towards grand nanotechnological visions revealed a desire for voice as well as choice. Interestingly, several of the scientists felt they were just as powerless as the lay participants in the face of economic and policy trends. They were refreshingly honest about their relative impotence in large, global innovation networks:

Sci   I mean scientists are in the same boat as everyone else I
        think. I mean scientists have no power to make these big
        choices. If you like, it's down to the politicians.

The public seemed to take reassurance from the open acknow-ledgements by the scientists of the conflicting pressures and constraints they routinely experienced. This seemed to humanise what 'nanoscience' involved for them, making it less threatening, but without resolving their fundamental concerns. For example:

> Pub   I think it was interesting that the scientists had the same
>   fears at the end of the day that we do. . . . So it's quite nice
>   at the end of the day we can relate to them as real people
>   and they do have the same fear.

This sense that 'we are all in it together' – was a central theme of the afternoon plenary discussion. That conversation also focused, repeatedly, on issues of responsibility and political agency, and the difficulty of ensuring the science was used for positive rather than negative ends:

> Pub   I think one thing that concerns me still and came out
>   further today is that I don't see who's really taking
>   responsibility for this. It's like scientists do science for
>   science's sake. Industry makes money for industry's sake.
>   Marketing departments market things because that's
>   their job but who is carrying the buck? That's the one
>   thing that I am most concerned about. Who at the end of
>   the day is responsible?
> Pub   I'd be interested to know if the scientists feel they do have
>   that responsibility, or if they should have it, or if they feel
>   they're kind of powerless.

What was striking in the discussion was the extent to which the scientists were as concerned, and frequently as perplexed, by such issues as the public. They were all citizens together on such matters. The scientists also saw themselves as being at the start of long chains of causation, involving investors, venture capital, competitive pressures, policy and economic forces. They seemed to believe that the imaginaries of others would be more influential than their own:

> Sci   It's only really today that I've realised I don't really feel that much responsibility at all. Because what I'm doing at the minute isn't actually used for anything, probably may not ever be used for anything, so I can't really take responsibility for what may happen in 20 years based on what I've done. Which may or may not be right.

A number of the scientists reflected on what responsibility means in relation to their work. What emerged was a picture of scientists making ethical decisions more or less in isolation:

> Sci   I think it's up to people to draw their own lines.
>
> Pub   If you were approached as an individual and you said no, they'd go somewhere else then, until they found a scientist who would work on their project.
>
> Int   The other point is that it's very hard to know what's dodgy and what isn't. So how do you work through that thinking process, which must involve thinking about how it's going to be used, if it can be used for other purposes? It's quite a complicated set of questions.
>
> Sci   It is. And it's entirely personal.

The role and responsibility of Albert Einstein in the development of nuclear weapons was a common touchstone in this discussion. Einstein was taken to represent the brilliant 'pure' scientist, who still could not ignore the context and implications of his work. Debate centred on whether Einstein's responsibility was personal – and therefore concerned only 'the science' – or collective – and involving wider considerations. The sense of individualised responsibility among the scientists was of some concern to the public. After the workshop, one explained:

Pub   Some of the thoughts I had about the scientists were reaffirmed, in that I don't think they look at the bigger picture. . . . And I think they should do. And I think the ethical side of it worries me as well. It seems to be down to the individual scientist as to whether they do something or not. But surely there should be some greater process?

But there also emerged a collective sense of the sheer complexity of contemporary scientific practice. Because of this complexity, and the influence of political, corporate and other interests on academic science, there were no easy answers to the question of who is responsible for nanotechnology. Towards the end of the discussion, there was a growing recognition that what is at stake in nanotechnology are the 'worlds we want to live in'. As such, the specific implications, benefits and risks of nanoscience pale in relation to the much larger political choices that we face:

Sci   The point I think I'd make is a lot of this stuff isn't actually about science. It's about politics in a broad sense. It's about what kind of world you want to live in 20 years from now. So I think it's better if we frame the question that way around: 'What are the possible futures?' and 'What kind of world do you want to live in?' From that you can build in how science [can] contribute and not contribute, and what are the dangers of it contributing to some undesirable future world.

Governing at the nanoscale

Beforehand, we were rather uncertain how the workshop would go. Would the scientists and the public participants argue or agree? Would they run out of interesting things to say? But afterwards, all the participants seemed to feel it had been a very positive experience. Each of them was interviewed at the end of the day and asked to give feedback. Most said that the workshop had helped them to recognise each other as fellow citizens with shared concerns. Both groups felt that there needed to be more opportunities for this type of conversation.

So at the end of the project, what wider conclusions can we draw? Elsewhere, we have discussed the need for institutional and policy innovations that can strengthen our collective capacity to grapple with the social dimensions of emerging technologies.[85] In particular, we have called for new institutional spaces (such as a commission for emerging technologies and society) and for changes to research and funding cultures that can make reflection and wider dialogue on social and ethical issues 'a normal and integral part' of the scientific process.[86] We will not rehearse those arguments again here. Instead, at a time when the quantity, diversity and international spread of public engagement processes around nanotechnologies is increasing,[87] let us draw out a few lessons for future processes.

First, the acknowledged complexity of governing nanotechnology emphasises the need for a wider range of actors to be drawn in processes of public engagement, particularly corporate scientists, strategists, venture capitalists and research funders. While many of the participants felt that the experience of taking part was enjoyable and in itself empowering, the sense of a lack of agency that came out of the discussion suggests that these other actors should be included in future public engagement exercises:

Sci If on the one hand there's a sense from the focus groups that people want to get involved in these discussions, I just feel they don't have any power to get involved and feel kind of left on the outside. . . . Most of the scientists said something quite similar. You know, they were interested in how their research was going to be used, but you're a PhD student, you can't get the bloody thing to work, you have no influence over how it's going to be used. So in a way, you can have this perfect dialogue between us, but nobody apparently has any power. I don't know if anybody else would disagree?

A second question is how to capitalise on the common concerns shared by both sets of participants. Several of the scientists expressed real surprise at the quality and intelligence of their exchanges with the public participants, and the amount of consensus that had emerged. Some even sought to distinguish *these* publics (informed as they were by two short focus groups) from 'normal' publics, who would be less sympathetic:

> Sci   A couple of us were saying that it's a very valuable process and obviously if we weren't all interested in communicating science to society we wouldn't be here. But the fact that the members of the public who are here today are already quite well informed is maybe making it not a reflective process. It's not something that you could just magnify. Because they want to be here, they already know quite a lot about nanotechnology. You couldn't transpose that to the public.

But the sense of surprise that these scientists feel is actually a valuable commodity in strengthening forms of scientific citizenship.[88] Challenging scientists' assumptions about the public can be an important step towards a more mature conversation about science and its ends.

Finally, there is clearly a need to distinguish between public engagement as a research exercise – providing a mechanism for evaluating the emergence of public opinion – and engagement that is intended to inform political, institutional and cultural change. Although the potential for public controversy provided a background rationale for this workshop, one limitation was that it was not designed to inform particular decisions or institutions. This is not to say it cannot have some kind of positive influence, either on the participants themselves, or through publications such as this. But our participants felt – and we would agree – that whenever possible, such processes should be designed in a way that connects to real decisions.

The box below details some further public engagement exercises that have grown out of the Lancaster–Demos project. These have a more direct link to decision-making. But it is through this project, and the opportunities it provided to experiment with new approaches, that we have developed our understanding of the theory and practice of upstream engagement. We will now begin to apply these insights elsewhere.

The Nanodialogues: experiments in public engagement

Over the course of 2006, with support from the government's Sciencewise programme, Demos and Lancaster University will be facilitating a series of practical experiments in public engagement, designed to inform decision-making around nanotechnologies.

Experiment 1: Nanoparticles and upstream regulation
Partner: the Environment Agency
Working with the Environment Agency, our first experiment explored how discussions between regulators and the public can contribute to sustainable innovation and regulation of nano-technologies. The Environment Agency is a firm advocate of 'risk-based regulation', but the uncertainties surrounding nanoparticles in complex ecosystems make risk assessment very difficult. Through a people's inquiry, which took place in January and February 2006, the Agency invited public inputs to emerging thinking about nanoparticles, regulation and environmental remediation.

Experiment 2: Imagining publicly engaged science
Partners: BBSRC and EPSRC
Research councils are a key influence on what is considered to be valuable science. The aim of our second experiment is to investigate the potential for public debate at an early stage in decision-making. Working with two of the research councils – Biotechnology and Biological Sciences, and Engineering and Physical Sciences – we will be exploring what might be at stake as

biotechnologies and nanotechnologies converge. How can dialogue between scientists and the public clarify key questions? And how can research priorities reflect public concerns?

Experiment 3: Nanotechnologies in development
Partner: Practical Action

Too often, the voices of people in developing countries are neither sought nor taken account of in decisions about science and innovation. For our third experiment, we are working with Practical Action, the development NGO, which has a lot of experience of public participation in developing countries. Practical Action will facilitate discussions with two community groups in southern Africa about the potential contribution of nanotechnologies to the provision of clean drinking water. Two UK nanoscientists will be invited to participate, as a way of deepening their understanding of local contexts, priorities and needs, and the implications these might have for their research.

A fourth experiment, in partnership with Unilever, is still in the process of being finalised, but will focus on public attitudes and engagement with corporate R&D. This will take place in autumn 2006.

Afterword

Brian Wynne

I have been given the task of reflecting on our role as analysts and actors – modest witnesses, as Donna Haraway aptly puts it[89] – in the (nano)science, technology and society debate. I also want to indicate some future possibilities for social science to contribute in a different way to these issues.

For some time at Lancaster University we have been attempting to conduct social scientific research which walks a tightrope between the worlds of academia and public policy. This approach has been further developed through our collaborations with Demos, which as a think tank is more obviously embedded in public policy. Our approach to researching science, technology and society has regularly required interventions in the routines of public and scientific institutions. Clarifying the nature of such engagements, and reflecting critically on them, is an essential safeguard against false or misdirected influence, or failing academic standards – or both.

In many ways, nanotechnologies present an extraordinary opportunity to build social science insight into the 'early stages' of the development of an emerging field. By taking the reflexive governance of nanotechnology as a central concern, social science has novel opportunities to become a modest actor in these changes, and to provide insights that are then co-produced with the scientific, technological and social changes they witness. We deliberately avoid the conventional characterisation of this as a social science *of* nanotechnology, precisely because this approach automatically establishes its object, 'nanotechnology', as if this were independent and well defined, rather than constantly evolving.

A new model of research practice

Long before the official shift away from the deficit model, research at Lancaster, and elsewhere, contradicted the prevailing account of publics and their ways of reacting to science and technologies. This contributed to the shift, marked by the House of Lords report of 2000,[90] towards something more open, two-way, exploratory and participatory. The commitment to 'public engagement with science' emerged, replacing the discredited but deeply entrenched 'public understanding of science' paradigm.

My own elaborations of this, from the 1990s onwards, distinguish between the need for ('upstream') public accountability of the assumed ends and purposes of scientific research, and public engagement only with the science of ('downstream') risks and other anticipated consequences. This reflected our own listening to ordinary citizens in qualitative fieldwork research situations: public meetings, structured focus group discussions, interviews, participant observation and so on.

During the period of this project's research into nanotechnology, this argument appears to have been accepted. A graphic example is the UK government's latest ten-year strategy for science and innovation, which includes a commitment 'to enable [public] debate to take place "upstream" in the scientific and technological development process, and not "downstream" where technologies are waiting to be exploited but may be held back by public scepticism brought about through poor engagement and dialogue on issues of concern'.[91] The same point has been made in relation to nanotechnologies by the Royal Society, Lord Sainsbury, the UK Science Minister, and the Office of Science and Technology,[92] as well as by the European Commission.

My first statement of this upstream logic, in Dublin in 1999,[93] was inspired by three main things:

○   a critical view of mainstream social science's reproduction
     of natural science's definitions of public concerns as being
     only about risks and consequences; the agenda for the
     social sciences was imagined to begin only when science
     was on the point of giving rise to its impacts – an
     entrenched downstream imagination reproduced across
     multiple disciplines

○   the relatively recent academic tradition of empirically
     grounded sociological research on scientific knowledge,
     which informed my concern that public experiences and
     concerns were being misconceived and misunderstood by
     looking only at *publics*, and not also at their relational
     'others' ('science', for example)

○   public respondents, who repeatedly asked about
     innovation-oriented as well as protection-oriented (risk)
     science; this prompted the 'upstream' shift, which sought
     to problematise some unacknowledged social and cultural
     dimensions of scientific knowledge.

It has been striking to see the rapid official uptake in UK and EU
science policy communities of the idea of upstream public
engagement, informed by analytical approaches showing how
technologies cannot be black-boxed and separated from sets of
constitutive social relations.[94] Such a conceptual approach lies in
stark contrast to the more confined role of the social sciences in the
development of biotechnologies and other domains. Dominant
approaches in a range of fields unquestioningly reproduced a framing
that assumed the emergent technology as a given, and thus assumed
the project of social reflection to be reduced to conceptualising,
assessing and managing 'impacts' only.

This powerful institutional framing of the social and ethical
dimensions of science has corresponded with an entrenched scientific
worldview which asserts that the social and ethical agenda only begins
after 'basic curiosity-driven science' has revealed the facts of nature's
possibilities. This linear model of science and practice is an iconic,

almost defining figuration of modernity. Yet although it does reflect some real driving motivations and commitments at the individual level, the putative boundary between basic science and its applications is deeply ambiguous. Social science cannot begin to understand public concerns until we also address what might be the objects of their concerns, including the social relationships in which these concerns are constituted. We were compelled to observe therefore that the social sciences risked becoming part of the problem, rather than an aid to potential resolution.[95]

The research set out in this pamphlet addresses this challenge in a practically engaged as well as theoretical way. This has been a constructive project, involving robust interactions with publics, scientists and technologists. Though not without creative tensions, our relationships with nanoscientists have been positive and collaborative. Our criticisms, where we have them, tend to be directed towards positivist social sciences, and institutional uses of these, rather than towards nanoscience as practised.

However, the more complex shift of focus, which the move 'upstream' was intended to introduce, has frequently been misunderstood. This can be noted, for example, in the otherwise admirable Royal Society/RAE report on nanotechnologies. This describes the potential role of upstream engagement in anticipating sensitive issues, despite our emphasis that upstream forms of public engagement with science are emphatically not about earlier *prediction* (and subsequent management) of impacts.

The idea of upstream engagement does not imagine that the aim is to provide earlier anticipation of impacts and risks (which is almost a contradiction in terms). Nor does it envisage publics telling scientists what to do or think. In any engagement process, the first people to acknowledge public deficits of knowledge are those publics themselves, and they are typically well aware of their unfitness to assume such a notionally active, direct influence on scientific work.

The roles of social scientists

So what are the appropriate roles of social scientists in 'upstream' debates around (nano)sciences and technologies? Taking a lead from discussions of 'public sociology' that are under way in the US,[96] we can see five distinct though overlapping roles:

O   studying public opinion and political mobilisation processes, with the hope that 'upstream' social intelligence helps to anticipate public controversy

O   helping to shape innovation processes and 'pick winners' by identifying wider public and consumer attitudes

O   enhancing public communication of (nano)science and technologies

O   eliciting public forms of knowledge that are relevant to identifying, assessing and managing risks

O   interpreting and reporting (thus, *representing*) public concerns and questions, which have not been recognised by policy experts.

The last two have been the principal framings of our own work. We have sought to open up the 'black boxes' of science and innovation, to induce greater reflexive awareness among scientists and others. In this way, innovation processes may *indirectly* gain added sensitivity to diverse human needs and aspirations, and so achieve greater resilience and sustainability.

Of course different actors, reflecting different powers and resources, will give different influence and scope to whichever of these roles they favour, or find themselves in. In this sense, the variegated field of nanotechnologies represents an opportunity to innovate more reflexive relationships between the social sciences, the physical sciences and policy. At the same time, it would be a mistake to take a simplistic or linear view of the 'upstream–downstream' metaphor and thereby view nano as a uniformly upstream issue – or indeed to see issues such as GM crops as monolithically downstream. The 'upstream' metaphor was never intended to be a catch-all model. But it made one central analytical point of distinction – the difference between *innovation*-oriented and *protection*-oriented science – from which much else follows.

However, despite the demand for further social science input, it remains true that institutionally dominant assumptions tend to leave 'upstream' questions over imagined social benefits and purposes to existing unquestioned institutional arrangements. This innocent enculturation is what needs to be disrupted. In the conventional model of public engagement, the expected role of the social sciences is tantamount to delivering a quiescent public for commercially exploitable scientific knowledge. What can be fleetingly glimpsed as a new approach reverts quickly to new versions of the supposedly dead-and-buried deficit model. 'Dialogue' once again becomes monologue.

A more realistic social science would, for the sake of robust public legitimacy, encourage policy to take seriously the challenges involved in reconciling the contradictory commitments which have been made: on one hand, investing in science for its competitive, knowledge-economy, wealth-creating purposes; and on the other, to cultivating meaningful forms of public engagement with science. This challenge, even in its academic sense, is not only intellectual, but also a challenge to the ethos of social science, its visions of what it should be for, how it relates to its users and sources of patronage, and how it defines its own objects of attention – 'society', 'the social' and 'the public'. Of course, this cannot be a singular ethos; but at least in the growing domain of public interactions with science, this vision of a more interpretive, engaged actor deserves a place.

What kind of social science?

The debate around nanotechnologies represents a novel opportunity for the social sciences to contribute to shaping scientific research trajectories. However, we should bear in mind that this terrain has traditionally been regarded as 'pure science', fenced off from social attention. How can it be approached in a way which helps science to flourish as an independent, cosmopolitan presence, but simultaneously opens it up to debate over its ends, expectations and imaginaries?

A key problem with the deficit model is the way that social science is itself encouraged to misconceive its own research objects, namely social actors. These misconceptions are then transmitted to its various users – who reinforce them in an innocent but vicious circle. As social scientists working on such issues, we should ask ourselves about the premises we bring to defining and conducting our own research and policy interventions.

If publics are thought to be responding to 'science' and its risks, and we are supposed to be understanding those public responses, then we must attempt to understand what that abstract conceptual object 'science' actually means for them. Usually this is taken for granted, and left untouched. What is it that they are responding to? What are the objects of their experience that give these responses saliency and meaning? And how might these indigenous meanings correspond, or conflict, with the meanings presumed by scientists, policy experts *and social scientists* onto those same publics?

These questions are endemically open, and not amenable to unambiguous answer, because citizens are *intrinsically* relational, not autonomous agents with autonomous 'values' and 'attitudes'. This is the premise which mistakenly shapes most social science and policy in this area.

Ambivalence, 'risk', (mis)trust and dependency

Mainstream social science, like the policy culture it influences, tends to take for granted the widely assumed model of people as autonomous individual subjects. Their relationship to science and technology is understood only in terms of its objects, perceived risks and benefits. Thus, social scientific accounts of UK 'public ambivalence' towards GM crops have defined such ambivalence as the conflict between perceptions of opposite objects – benefits and risks.[97]

I articulated an alternative perspective, more than a decade ago, when I suggested that such public reactions, including ambivalence, were always *relational* experiences, and that risk itself was a fundamentally relational condition.[98] Thus 'risk' as known potential harm always carries in its shadow the prospect of unknown effects, which by definition cannot be expressed in any risk assessment. Field research has shown repeatedly that public awareness of expert ignorance and the likelihood of unpredicted consequences is a central element of public concern in the face of risky technologies. This is a central, not contextual, social–relational concern about the institutions on which we are all dependent in such matters.

So ambivalence is structured not in terms of risks and benefits, but in the very different terms of *social–institutional relations*. Publics are typically ambivalent because they know they are unavoidably dependent on institutions of science, expertise and policy which they do not trust, and which exaggerate their own knowledge and control.

A key condition for rebuilding trust and legitimacy in such institutions is that they must be experienced to be putting their own assumptions into dialogue with others. This deepening of the scope of interaction is critical if they are wanting any kind of meaningful dialogue.[99]

Circles of ambiguity – and modest development?

The mode of social science presented here involves more than intellectual dimensions alone. It also involves learning new relationships and responsibilities, with 'the public', with the natural sciences and with policy. And it involves social sciences becoming actors in those worlds as well as commentators.

However, this leaves a continuing issue unresolved. If we are to engage in these more politically immersed relationships, and leave behind our well-bounded peer cultures, how are we to ensure that the knowledge we generate can claim validity? If we are interpreting and constructing representations of public concerns and ways of reasoning, and then acting on these in public arenas, are we claiming grandiosely to 'know what people think'?

In one sense our open-ended fieldwork, which allows public respondents to define their own meanings and not to be forced into the assumed meanings that are built into opinion surveys, does still claim a positivist representational status. When we find that people say, for example, that they are concerned more about unknown effects than known risks, and about the apparent denial of these by institutionalised science, we say that this is a 'typical' public attitude about such issues. If we then present these findings to scientists and policy-makers, we are representing 'publics'. But we are doing this in a questioning, exploratory and tentative way. The intention is not to advocate any partisan stance, but to invite self-reflection on the part of the institutional actors about whom these public observations have been made. We are saying, in effect: 'Let's imagine that this is a genuine public concern, representative or minority, it does not matter. *What is our responsibility in relation to this?'*

Extending this ambiguity requires a combination of academic analysis and further 'testing' in the domain of public debate itself. We have to ask: 'Do these intellectual constructions resonate or not with evolving public attitudes and responses?' There is a normative dimension to this work, for which we take responsibility; but it is one which avoids partisan positions, even if it occasionally finds itself lined up uncomfortably close to one or another of these. Rather, it addresses a different normative issue: the development of citizens, society and scientific cultures as nothing more and nothing less than a 'modest witness', asking pertinent questions and inviting pertinent reflection.

Brian Wynne is Professor of Science Studies at Lancaster University and a senior partner in the ESRC Centre for the Economic and Social Aspects of Genomics (CESAGen).

Notes

1   House of Commons Science and Technology Committee, *Too Little Too Late? Government investment in nanotechnology*, Fifth Report of Session 2003–04 (London: House of Commons, 2004); National Science Foundation, *Societal Implications of Nanoscience and Nanotechnology* (Arlington VA: NSF, 2001).

2   Cientifica, *Where Has My Money Gone? Government nanotechnology funding and the $18 billion pair of pants* (London: Cientifica, 2006); available at www.cientifica.com/www/details.php?id=340 (accessed 20 Mar 2006).

3   Editorial, 'Nanotechnology is not so scary', *Nature* 421, no 6921 (2003); Royal Society and Royal Academy of Engineering, *Nanoscience and Nanotechnologies: Opportunities and uncertainties* (London: Royal Society/RAE, 2004).

4   Royal Society/RAE, *Nanoscience and Nanotechnologies*.

5   Interviews with Professor Raymond Baker (former CEO, Biotechnology and Biological Sciences Research Council, BBSRC); Professor John Beringer (former chairman, Advisory Committee for Releases to the Environment, ACRE); Sir Thomas Blundell (former CEO, BBSRC); Mark Cantley (adviser, Directorate for Life Sciences, European Commission); Dr Ian Gibson MP (former chairman, House of Commons Science and Technology Select Committee); Julie Hill (former ACRE member and director, Green Alliance); Professor Sir Martin Holdgate (former chief scientist, Department of the Environment); Dr Sue Mayer (director, Genewatch); Dr Doug Parr (chief scientist, Greenpeace); and Professor Nigel Poole (former chief bioscientist, Zeneca). All these interviews took place from February to April 2004.

6   A Hedgecoe and P Martin, 'The drugs don't work: expectations and the shaping of pharmacogenetics', *Social Studies of Science* 33, no 2 (2003).

7   G Marcus (ed), *Technoscientific Imaginaries* (Chicago: University of Chicago Press, 1995).

8   Interviews with Rainer Gerold and Nicole Dewandre (Science and Society Directorate, European Commission); Clive Hayter (Engineering and Physical

Sciences Research Council, EPSRC); Mark Modzelewski (Lux Research/NanoBusiness Alliance); Professor Philip Moriarty (Nottingham University); Christine Peterson (Foresight Institute); David Rejeski (Woodrow Wilson Center); Professor George Smith (Oxford University); Jim Thomas (ETC group); and Renzo Tomellini (Nanoscience and Nanotechnologies Unit, European Commission).

9   See, for example, D Taverne, *The March of Unreason* (London: Oxford University Press, 2005).

10   A Nordmann, 'Molecular disjunctions: staking claims at the nanoscale' in D Baird, A Nordmann and J Schummer, *Discovering the Nanoscale* (Amsterdam: IOS Press, 2004). ·

11   See, for example, EF Einsiedel and L Goldenberg, 'Dwarfing the social? Nanotechnology lessons from the biotechnology front', *Bulletin of Science, Technology & Society* 24 (2004); S Mayer, 'From genetic modification to nanotechnology: the dangers of "sound science"' in T Gilland (ed), *Science: Can we trust the experts?* (London: Hodder and Stoughton, 2002); MD Mehta, 'From biotechnology to nanotechnology: what can we learn from earlier technologies?', *Bulletin of Science, Technology & Society* 24 (2004); JR Wolfson, 'Social and ethical issues in nanotechnology: lessons from biotechnology and other high technologies', *Biotechnology Law Report* 22, no 4 (2003).

12   R Sandler and WD Kay, 'The GMO–nanotech (dis)analogy?', *Bulletin of Science, Technology & Society* 26, no 1 (2006).

13   J Wilsdon and R Willis, *See-through Science: Why public engagement needs to move upstream* (London: Demos, 2004); P Macnaghten, M Kearnes and B Wynne, 'Nanotechnology, governance and public deliberation: what role for the social sciences?', *Science Communication* 27, no 2 (2005).

14   The principal focus of our analysis is on the 1980s and 1990s, up to the moment when the controversies over the first period of GM development reached their peak in the UK, in February 1999. Clearly, since that time, there have been a number of further developments, including the creation of the Agriculture and Environment Biotechnology Commission (AEBC), the UK government's GM dialogue, completion of the farm-scale trials, and, not least, the hearings at the World Trade Organization (WTO) into the formal US complaint on 'biotech products' against the EU. But we have drawn the line at February 1999, in order to reflect on the underlying processes which shaped the controversies, rather than the unfolding of the post-1998 events themselves.

15   CH Waddington, *The Man-Made Future* (London: Croom Helm, 1978).

16   Interview with Professor Nigel Poole, former chief bioscientist, Zeneca, 16 Mar 2004.

17   R Doubleday, 'Political innovation: corporate engagements in controversy over genetically modified foods', unpublished PhD thesis (London: University College London, 2004).

18   B Wynne, 'Public understanding of science' in S Jasanoff et al (eds), *Handbook of Science and Technology Studies* (Thousand Oaks, CA: Sage,1995).

19   J Ravetz, 'The post-normal science of safety' in M Leach, I Scoones and B
     Wynne, *Science and Citizens: Globalization and the challenge of engagement*
     (London: Zed Books, 2005).
20   S Mayer and A Stirling, 'Finding a precautionary approach to technological
     developments – lessons for the evaluation of GM crops', *Journal of Agricultural
     and Environmental Ethics* 15 (2002).
21   J Tait and L Levidow, 'Proactive and reactive approaches to regulation: the case
     of biotechnology', *Futures* 24, no 3 (1992).
22   Interview with Professor John Beringer, 23 Mar 2004.
23   Ibid.
24   S Jasanoff, *Designs on Nature: Science and democracy in Europe and the United
     States* (Princeton: Princeton University Press, 2005).
25   R Grove-White et al, *Uncertain World: Genetically modified organisms, food and
     public attitudes in Britain* (Lancaster: Institute for Environment, Philosophy
     and Public Policy (IEPPP), Lancaster University, 1997); R Grove-White, P
     Macnaghten and B Wynne, *Wising Up: The public and new technologies*
     (Lancaster, UK: IEPPP, Lancaster University, 2000).
26   G Gaskell et al, 'GM foods and the misperception of risk perception', *Risk
     Analysis* 24, no 1 (2004).
27   S Joss and J Durant, *Public Participation in Science: The role of consensus
     conferences in Europe* (London: Science Museum, 1995); R Macrory, *National
     Biotechnology Conference: Report of the rapporteur* (London: Department for
     Environment, Transport and the Regions, 1997).
28   B Wynne, *Rationality and Ritual: The Windscale inquiry and nuclear decisions in
     Britain* (Chalfont St Giles: British Society for the History of Science, 1992).
29   J Durant, M Bauer and G Gaskell, *Biotechnology in the Public Sphere: A
     European source book* (London: Science Museum, 1998); National Institute for
     Agricultural Research (INRA), *Biotechnology and Genetic Engineering: What
     Europeans think about it in 1993*, Eurobarometer 39.1 (Brussels: European
     Commission, 1993); INRA, *The Europeans and Biotechnology*, Eurobarometer
     52.1 (Brussels: European Commission, 2000); MORI, *The Public Consultation
     on Developments in the Biosciences* (London: Department of Trade and
     Industry, 1999).
30   House of Lords Select Committee on Science and Technology, *Science and
     Society* (London: House of Lords, 23 Feb 2000).
31   Interview with Professor Raymond Baker, former CEO, BBSRC, 24 Feb 2004.
32   Interview with Professor Nigel Poole.
33   Indeed, as a response to the perception that such groups were not campaigning
     actively on genetically modified organisms (GMOs) from the mid 1990s, wider
     bodies of opinion, independent of such organisations, crystallised in a host of
     more ad hoc and GM-specific networks – including Genetix Snowball, the
     Genetics Network, the Genetics Alliance, Corporate Watch, Genewatch and
     many others. This further range of frequently internet-focused associations
     embraced wide and diverse constituencies of concern, and can be read as
     'organisational' crystallisations of the pervasive, but previously latent, public

unease about GM-related issues noted in UK social research as early as 1996–97 (Grove-White et al, *Uncertain World*).

34 Interview with Dr Doug Parr, chief scientist, Greenpeace, 4 Mar 2004.

35 B Wynne, 'Creating public alienation: expert cultures of risk and ethics on GMOs', *Science as Culture* 10, no 4 (2001); Jasanoff, *Designs on Nature*.

36 B Latour, 'Give me a laboratory and I will raise the world' in KD Knorr-Cetina and MJ Mulkay (eds), *Science Observed* (Beverly Hills: Sage, 1983).

37 B Latour, *Science in Action: How to follow scientists and engineers through society* (Cambridge, MA: Harvard University Press, 1987).

38 This research forms the empirical basis of this chapter. In publishing this material we have sought to protect the anonymity of the nanoscience researchers we interviewed in Cambridge (September and October 2004) and Oxford (December 2004). As such we refer to such scientists simply as 'researchers', though through the course of our research we interviewed scientists at every level of the scientific career structure.

39 This feature of nanoscience research has its roots in post Second World War 'big science' research efforts – particularly associated with the development of nuclear technologies, space exploration and high-speed physics. See, for example, P Galison and B Hevly, *Big Science: The growth of large-scale research* (Stanford, CA: Stanford University Press, 1992).

40 See www.nanoscience.cam.ac.uk/irc/pdf/irc_nanoscience_proposal.pdf (accessed 20 Mar 2006).

41 A Rip, 'Folk theories about nanotechnology', *Science as Culture* (2006, forthcoming).

42 See www.nanoscience.cam.ac.uk/irc/pdf/irc_nanoscience_proposal.pdf (accessed 20 Mar 2006).

43 D Baird, *Thing Knowledge: A philosophy of scientific instruments* (Berkeley: University of California Press, 2004); D Baird and A Shew, 'Probing the history of scanning tunnelling microscopy' in D Baird, A Nordmann and J Schummer (eds), *Discovering the Nanoscale* (Amsterdam: IOS Press, 2004); CCM Mody, 'How probe microscoptists became nanotechnologists' in D Baird et al, *Discovering the Nanoscale*.

44 N Taniguchi, 'On the basic concept of "nano-technology"', *Proceedings of the International Conference of Production Engineering* (Tokyo: Japan Society of Precision Engineering, 1974).

45 KE Drexler, *Engines of Creation: The coming era of nanotechnology* (New York: Anchor Books, 1986).

46 Interview with Philip Moriarty, 31 Mar 2005.

47 Interview with researcher, Cambridge NanoScience Centre, autumn 2004.

48 Interview with researcher, Department of Material Science, Oxford University, Dec 2004.

49 Interview with Christine Peterson, 21 Mar 2005.

50 See, for example, R Smalley, 'Of chemistry, love and nanobots', *Scientific American* 285 (2001); R Smalley, 'Smalley responds', *Chemical & Engineering*

News 81 (2003); R Smalley, 'Smalley concludes', *Chemical & Engineering News* 81 (2003).

51   Interview with Nadrian Seeman, 12 Apr 2005.

52   Bernadette Bensaude-Vincent suggests that there are 'two cultures of nanotechnology'. While one culture uses a mechanistic approach of directly controlling atoms and molecules, biomimetic nanotechnology is characterised by an attempt to model existing biological systems. See B Bensaude-Vincent, 'Two cultures of nanotechnology?', *Hyle* 10 (2004).

53   Interview with researcher, Cambridge NanoScience Centre, autumn 2004.

54   Interview with researcher, Department of Biophysics, Cambridge University, autumn 2004.

55   National Science and Technology Council (NSTC) Interagency Working Group on Nanoscience, Engineering and Technology (IWGN), *National Nanotechnology Initiative: Leading to the next industrial revolution* (Washington DC: NSTC, 2000); available at www.ostp.gov/NSTC/html/iwgn/iwgn.fy01budsuppl/toc.htm (accessed 20 Mar 2006).

56   MC Roco and WS Bainbridge (eds), *Converging Technologies for Improving Human Performance: Nanotechnology, biotechnology, information technology and the cognitive science* (Arlington, VA: National Science Foundation, 2002).

57   Department of Trade and Industry, *New Dimensions of Manufacturing: A UK strategy for nanotechnology* (London: DTI, 2002).

58   O Saxl, *Nanotechnology IS Important* (Institute for Nanotechnology, 2002); available at www.iee.org/oncomms/pn/materials/NANOTECHNOLOGY_REPORT.pdf (accessed 20 Mar 2006).

59   DTI, *New Dimensions of Manufacturing.*

60   Interview with researcher, Department of Material Science, University of Oxford, Dec 2004.

61   Interview with researcher, Cambridge Nanoscience Centre, autumn 2004.

62   DTI, *New Dimensions of Manufacturing.*

63   Interview with Renzo Tomellini, 17 Jun 2004.

64   Interview with researcher, Cambridge Nanoscience Centre, autumn 2004.

65   Interview with Mark Welland, director, Cambridge Nanoscience Centre, 1 Oct 2004.

66   NSTC, *National Nanotechnology Initiative.*

67   Royal Society/RAE, *Nanoscience and Nanotechnologies.*

68   Interview with researcher, Department of Material Science, Oxford University, Dec 2004.

69   Interview with researcher, Department of Material Science, University of Oxford, Dec 2004.

70   G Gaskell, N Allum and S Stares, *Europeans and Biotechnology in 2002*, Eurobarometer 58.0 (London: Methodology Institute, London School of Economics, 2003).

71   Royal Society/RAE, *Nanoscience and Nanotechnologies.*

72   M Cobb and J Macoubrie, 'Public perceptions about nanotechnology: risks, benefits and trust', *Journal of Nanoparticle Research* 6 (2004); J Macoubrie, *Informed Public Perceptions of Nanotechnology and Trust in Government* (Washington, DC: Woodrow Wilson International Center for Scholars, 2005), available at: www.pewtrusts.com/pdf/Nanotech_0905.pdf (accessed 20 Mar 2006).

73   In the US, 80% of survey respondents indicated they had heard little or nothing about nanotechnology (Cobb and Macoubrie, 'Public perceptions about nanotechnology') while in the UK, only 29% of respondents said they were aware of the term.

74   Cobb and Macoubrie, 'Public perceptions about nanotechnology'.

75   Macoubrie, *Informed Public Perceptions of Nanotechnology and Trust in Government*; Royal Society/RAE, *Nanoscience and Nanotechnologies*.

76   Cobb and Macoubrie, 'Public perceptions about nanotechnology'; Macoubrie, *Informed Public Perceptions of Nanotechnology and Trust in Government*; Royal Society/RAE, *Nanoscience and Nanotechnologies*.

77   See, for example, the Woodrow Wilson Center's inventory of nanotechnology in consumer products at www.nanotechproject.org/consumerproducts (accessed 20 Mar 2006).

78   Grove-White et al, *Uncertain World*; Grove-White et al, *Wising Up*.

79   On focus groups methods see M Bloor et al, *Focus Groups in Social Research* (London: Sage, 2001); and P Macnaghten and G Myers, 'Focus groups: the moderator's view and the analyst's view' in G Gobo et al (eds), *Qualitative Research Practice* (London: Sage, 2004).

80   M = male participant; F = female participant; Int = interviewer. The composition of the group is mentioned at the end of each quote.

81   T Zeldin, *Conversation: How talk can change our lives* (London: Harvill Press, 1998).

82   For details please refer to the event website http://demosgreenhouse.co.uk/mediawiki/index.php/Demos_Nano_Talk (accessed 20 Mar 2006).

83   Sci = nanoscientist; Pub = public participant.

84   For similar efforts in upstream public engagement see the 'Meeting of minds: European citizens' deliberation on brain science' project, www.meetingmindseurope.com/ (accessed 20 Mar 2006); the 'Nanodialogues: four experiments in upstream public engagement' project, www.demos.co.uk/projects/currentprojects/nanodialogues/ (accessed 20 Mar 2006); and 'Deliberative mapping: citizens and specialists informing decisions on organ transplant options' project, www.deliberative-mapping.org/ (accessed 20 Mar 2006).

85   See, for example, Wilsdon and Willis, *See-through Science*; Macnaghten et al, 'Nanotechnology, governance and public deliberation'; and J Wilsdon, B Wynne and J Stilgoe, *The Public Value of Science: Or how to ensure that science really matters* (London: Demos, 2005).

86   This phrase is taken from the House of Lords Select Committee on Science and Technology, *Science and Society*.

87   See, for example, the summary of activities in 'Policy Report 1' (Mar 2006) of the UK Nanotechnology Engagement Group; see www.involving.org (accessed 20 Mar 2006).

88   KG Davies and J Wolf-Phillips, 'Scientific citizenship and good governance: implications for biotechnology', *Trends in Biotechnology* 24, no 2 (2006).

89   D Haraway, *Modest_Witness@Second_Millennium.FemaleMan_Meets_Oncomouse™: Feminism and technoscience* (New York: Routledge, 1997).

90   House of Lords Select Committee on Science and Technology, *Science and Society*.

91   HM Treasury, Department of Trade and Industry, and Department of Education and Skills, *Science and Innovation Investment Framework 2004–2014* (London: HM Treasury, 2004).

92   Department of Trade and Industry (DTI), 'Nanotechnology offers potential to bring jobs, investment and prosperity', Lord Sainsbury, press release from the DTI, 29 Jul 2004; (DTI)/Office of Science and Technology (OST), 'OST grant scheme – Sciencewise: Engaging society with science and technology', 2005; available from www.sciencewise.org.uk/ (accessed 20 Mar 2006); Royal Society/RAE, *Nanoscience and Nanotechnologies*.

93   I first argued explicitly for 'upstream public engagement' in my keynote address, 'Acceptance of novel technologies', 7–9 April 1999, at the EU Federation of Biotechnology Workshop in Dublin.

94   See, for example, B Wynne, 'Unruly technology: practical rules, impractical discourses and public understanding', *Social Studies of Science* 18, no 1 (1988); TJ Pinch and WE Bijker, 'The social construction of facts and artefacts: or how the sociology of science and the sociology of technology might benefit each other', *Social Studies of Science* 14 (1984); J Law and J Hassard (eds), *Actor Network Theory and After* (Oxford: Blackwell, 1999).

95   Grove-White et al, *Wising Up*.

96   M Burawoy, 'A public sociology for human rights', introduction to J Blau and K Lyall Smith (eds), *Public Sociologies Reader* (Lanham, MD: Rowman and Littlefield, 2006); M Burawoy, 'Forging public sociologies on national, regional, and global terrains', *E-Bulletin of the International Sociological Association* 2 (2005) at www.ucm.es/info/isa/publ/e_bulletin.htm (accessed 24 Mar 2006); M Burawoy, 'For public sociology', address to the American Sociological Association, San Francisco, 15 Aug 2004, *American Sociological Review* (Feb 2005).

97   This was the definition and approach given in an analysis of the 2003 UK GM Nation debate by the University of East Anglia risk team, Nick Pidgeon and coworkers, at the CESAGen Genomics and Society conference, Apr 2004, at the London Royal Society.

98   B Wynne, 'Public Understanding of Science'.

99   B Wynne, 'Public engagement as a means of restoring trust in science? Hitting the notes, but missing the music', *Community Genetics* 10, no 5 (2006).

DEMOS – Licence to Publish

1. **Definitions**
   a **"Collective Work"** means a work, such as a periodical issue, anthology or encyclopedia, in which the Work in its entirety in unmodified form, along with a number of other contributions, constituting separate and independent works in themselves, are assembled into a collective whole. A work that constitutes a Collective Work will not be considered a Derivative Work (as defined below) for the purposes of this Licence.
   b **"Derivative Work"** means a work based upon the Work or upon the Work and other pre-existing works, such as a musical arrangement, dramatization, fictionalization, motion picture version, sound recording, art reproduction, abridgment, condensation, or any other form in which the Work may be recast, transformed, or adapted, except that a work that constitutes a Collective Work or a translation from English into another language will not be considered a Derivative Work for the purpose of this Licence.
   c **"Licensor"** means the individual or entity that offers the Work under the terms of this Licence.
   d **"Original Author"** means the individual or entity who created the Work.
   e **"Work"** means the copyrightable work of authorship offered under the terms of this Licence.
   f **"You"** means an individual or entity exercising rights under this Licence who has not previously violated the terms of this Licence with respect to the Work, or who has received express permission from DEMOS to exercise rights under this Licence despite a previous violation.
2. **Fair Use Rights.** Nothing in this licence is intended to reduce, limit, or restrict any rights arising from fair use, first sale or other limitations on the exclusive rights of the copyright owner under copyright law or other applicable laws.
3. **Licence Grant.** Subject to the terms and conditions of this Licence, Licensor hereby grants You a worldwide, royalty-free, non-exclusive, perpetual (for the duration of the applicable copyright) licence to exercise the rights in the Work as stated below:
   a to reproduce the Work, to incorporate the Work into one or more Collective Works, and to reproduce the Work as incorporated in the Collective Works;
   b to distribute copies or phonorecords of, display publicly, perform publicly, and perform publicly by means of a digital audio transmission the Work including as incorporated in Collective Works;
   The above rights may be exercised in all media and formats whether now known or hereafter devised. The above rights include the right to make such modifications as are technically necessary to exercise the rights in other media and formats. All rights not expressly granted by Licensor are hereby reserved.
4. **Restrictions.** The licence granted in Section 3 above is expressly made subject to and limited by the following restrictions:
   a You may distribute, publicly display, publicly perform, or publicly digitally perform the Work only under the terms of this Licence, and You must include a copy of, or the Uniform Resource Identifier for, this Licence with every copy or phonorecord of the Work You distribute, publicly display, publicly perform, or publicly digitally perform. You may not offer or impose any terms on the Work that alter or restrict the terms of this Licence or the recipients' exercise of the rights granted hereunder. You may not sublicense the Work. You must keep intact all notices that refer to this Licence and to the disclaimer of warranties. You may not distribute, publicly display, publicly perform, or publicly digitally perform the Work with any technological measures that control access or use of the Work in a manner inconsistent with the terms of this Licence Agreement. The above applies to the Work as incorporated in a Collective Work, but this does not require the Collective Work apart from the Work itself to be made subject to the terms of this Licence. If You create a Collective Work, upon notice from any Licencor You must, to the extent practicable, remove from the Collective Work any reference to such Licensor or the Original Author, as requested.
   b You may not exercise any of the rights granted to You in Section 3 above in any manner that is primarily intended for or directed toward commercial advantage or private monetary

compensation. The exchange of the Work for other copyrighted works by means of digital file-sharing or otherwise shall not be considered to be intended for or directed toward commercial advantage or private monetary compensation, provided there is no payment of any monetary compensation in connection with the exchange of copyrighted works.

c If you distribute, publicly display, publicly perform, or publicly digitally perform the Work or any Collective Works, You must keep intact all copyright notices for the Work and give the Original Author credit reasonable to the medium or means You are utilizing by conveying the name (or pseudonym if applicable) of the Original Author if supplied; the title of the Work if supplied. Such credit may be implemented in any reasonable manner; provided, however, that in the case of a Collective Work, at a minimum such credit will appear where any other comparable authorship credit appears and in a manner at least as prominent as such other comparable authorship credit.

5. Representations, Warranties and Disclaimer

a By offering the Work for public release under this Licence, Licensor represents and warrants that, to the best of Licensor's knowledge after reasonable inquiry:

i Licensor has secured all rights in the Work necessary to grant the licence rights hereunder and to permit the lawful exercise of the rights granted hereunder without You having any obligation to pay any royalties, compulsory licence fees, residuals or any other payments;

ii The Work does not infringe the copyright, trademark, publicity rights, common law rights or any other right of any third party or constitute defamation, invasion of privacy or other tortious injury to any third party.

b EXCEPT AS EXPRESSLY STATED IN THIS LICENCE OR OTHERWISE AGREED IN WRITING OR REQUIRED BY APPLICABLE LAW, THE WORK IS LICENCED ON AN "AS IS" BASIS, WITHOUT WARRANTIES OF ANY KIND, EITHER EXPRESS OR IMPLIED INCLUDING, WITHOUT LIMITATION, ANY WARRANTIES REGARDING THE CONTENTS OR ACCURACY OF THE WORK.

6. Limitation on Liability. EXCEPT TO THE EXTENT REQUIRED BY APPLICABLE LAW, AND EXCEPT FOR DAMAGES ARISING FROM LIABILITY TO A THIRD PARTY RESULTING FROM BREACH OF THE WARRANTIES IN SECTION 5, IN NO EVENT WILL LICENSOR BE LIABLE TO YOU ON ANY LEGAL THEORY FOR ANY SPECIAL, INCIDENTAL, CONSEQUENTIAL, PUNITIVE OR EXEMPLARY DAMAGES ARISING OUT OF THIS LICENCE OR THE USE OF THE WORK, EVEN IF LICENSOR HAS BEEN ADVISED OF THE POSSIBILITY OF SUCH DAMAGES.

7. Termination

a This Licence and the rights granted hereunder will terminate automatically upon any breach by You of the terms of this Licence. Individuals or entities who have received Collective Works from You under this Licence, however, will not have their licences terminated provided such individuals or entities remain in full compliance with those licences. Sections 1, 2, 5, 6, 7, and 8 will survive any termination of this Licence.

b Subject to the above terms and conditions, the licence granted here is perpetual (for the duration of the applicable copyright in the Work). Notwithstanding the above, Licensor reserves the right to release the Work under different licence terms or to stop distributing the Work at any time; provided, however that any such election will not serve to withdraw this Licence (or any other licence that has been, or is required to be, granted under the terms of this Licence), and this Licence will continue in full force and effect unless terminated as stated above.

8. Miscellaneous

a Each time You distribute or publicly digitally perform the Work or a Collective Work, DEMOS offers to the recipient a licence to the Work on the same terms and conditions as the licence granted to You under this Licence.

b If any provision of this Licence is invalid or unenforceable under applicable law, it shall not affect the validity or enforceability of the remainder of the terms of this Licence, and without further action by the parties to this agreement, such provision shall be reformed to the minimum extent necessary to make such provision valid and enforceable.

c No term or provision of this Licence shall be deemed waived and no breach consented to unless such waiver or consent shall be in writing and signed by the party to be charged with such waiver or consent.

d This Licence constitutes the entire agreement between the parties with respect to the Work licensed here. There are no understandings, agreements or representations with respect to the Work not specified here. Licensor shall not be bound by any additional provisions that may appear in any communication from You. This Licence may not be modified without the mutual written agreement of DEMOS and You.